BOLIVAR
Liberator of a Continent

BOLIVAR

Liberator of a Continent

A DRAMATIZED BIOGRAPHY

by

bill boyd

S.P.I. BOOKS
New York

BOLIVAR, Copyright © 1998 by Bill Boyd. All rights reserved. Printed in the United States of America. No part of this book may be used or reproduced in any manner without written permission of the publisher, except in the case of brief quotations embodied in critical articles or reviews. For further information, contact S.P.I. Books, 136 West 22nd Street, New York, NY 10011. Tel: 212/633-2023, Fax 212/633-2123

ISBN: 1-56171-944-3

First Edition September 1998

AUTHOR'S NOTE

Anyone who has ever been to South America knows that in every city and town there is a Plaza Bolivar, an Avenida Bolivar and, probably, a Hotel Bolivar. In Caracas, the airport is named after Simon Bolivar. And there are statues of Bolivar, the "Liberator," in almost every plaza that bears his name. However, it was a statue of Simon Bolivar at the south entrance to Central Park in New York City, which made me wonder, "Who exactly was this Simon Bolivar?" Even though I'm from Panama, a "Bolivarian" country, I knew only that he had fought the Spanish and gained the independence of most of South America, that he wore a uniform, rode a horse, and presumably led troops in battle. But why is he such an *exceptional* hero? The idol of Latin America? The demigod of an entire continent? At that point, I didn't know, but I knew I must find out.

During my research, I unearthed enough information to write a fifty-volume biography. The challenge was to portray the character of the man, his actions and his thoughts, his accomplishments and his failures in a few hundred pages. There was only one way: Eliminate everything extraneous and recount in detail the important episodes. Keep it factual without making it dull. BOLIVAR is a biography of the great South American freedom fighter. Although dramatized, the events depicted took place as described. With a few inconsequential exceptions, all of the individuals in the biography actually existed and their personalities and actions are derived from historical documents and the official archives of the period.

CHRONOLOGY

1783   Simon Bolivar born in Caracas, Venezuela, on July 24th.

1789   French Revolution.

1799   George Washington dies.

1801   Simon Bolivar goes to Spain to be educated and obtain "polish."

1802   Bolivar marries his cousin, Maria Teresa de Toro, in Madrid. She dies eight months later in Caracas.

1808   Joseph Bonaparte crowned King of Spain by his brother, Napoleon, deposing Ferdinand VII. Spanish colonies refuse to accept Joseph as king.

1810   Bolivar goes to England as Venezuelan envoy. Brings Francisco Miranda back with him to Venezuela.

1811   Venezuela declares independence on July 5th. Miranda placed at head of government.

1812   Earthquake in Caracas kills thousands. Crushes republican government.

Bolivar continues fighting in Venezuela until his defeat in 1814.

1815   British defeat Napoleon at the Battle of Waterloo.

Ferdinand VII returned to Spanish throne.

Americans defeat British at the Battle of New Orleans.

Bolivar in exile in Jamaica, then travels to Haiti. (At this time, Haiti and the United States are the only countries in the Western Hemisphere to have achieved independence.)

1816   Bolivar and 250 of his men return to Venezuela with money and arms donated by President Petion of Haiti.

1819   Bolivar convenes a Congress at Angostura, declaring he will rule by law under a constitution.

In one of history's greatest feats, Bolivar crosses the Andes, defeats Spanish at Boyacá, liberates Colombia.

1820   Armistice signed between Bolivar and Marshal Pablo Morillo. Morillo returns to Spain.

1821   Armistice broken. Outnumbered, Bolivar defeats Spanish Army at the Battle of Carabobo. All Venezuela is free.

With Colombia, they form the Republic of Great Colombia with Simon Bolivar as president.

Lima falls to General Jose de San Martin of Argentina.

Panama revolts from Spain, joins Bolivar's Great Colombia.

1822   Bloody battles of Bomboná and Pichincha liberate Quito.

Manuela Sáenz becomes Bolivar's mistress.

Bolivar annexes Guayaquil to Great Colombia.

Historic meeting between Bolivar and San Martin in Guayaquil.

1823	President Monroe issues Monroe Doctrine.
	Simon Bolivar enters Lima.
1824	Battle of Junin, Bolivar defeats Royal Cavalry.
	Battle of Ayacucho. Marshal Sucre defeats two Spanish Armies. All South America is free.
1827	Bolivar returns to Bogota. Conflict with Santander.
	Bolivar subdues Paez in Venezuela.
1828	Assassination plot against Bolivar. Manuela saves his life. However, the fact that some of the very people he freed tried to kill him affects the Liberator profoundly.
1830	Exiled, in Santa Marta, Colombia, Simon Bolivar dies of tuberculosis complicated by pneumonia on December 17th.
1842	Realization of Bolivar's genius and appreciation of his accomplishments cause a complete re-evaluation of the man. His body is brought back to Caracas to be placed in the Pantheon of Heroes, where he lies today, honored and revered as the father of six South American nations.

Chapter 1

It was early morning on May 8, 1830. It had rained the night before, and a chill hung in the air. In the courtyard of a small but beautiful villa on the outskirts of Bogota, the members of the cabinet, the leading citizens and the entire diplomatic corps had assembled on horseback. When Simon Bolivar emerged from the house, a murmur rippled through the crowd. Although he was only forty-seven years old, the great Liberator looked much older. He was frail, and his body, which was never any taller than five feet, six inches, seemed to have shrunk from his years of hard campaigning. In contrast to the bright uniforms and gold braid worn by many of the officers present, Bolivar was dressed in plain dark civilian attire. His only decoration was George Washington's Yorktown Medal, which he always wore around his neck. Jose Palacios, Bolivar's huge black companion who had fought at his side in more than two hundred battles, held the Liberator's horse while Bolivar put his foot into the stirrup. The leather creaked as he swung himself painfully into the saddle. The stern-faced, ever loyal General Urdaneta approached and said in a low voice, "The army is with you to a man, sir. All you have to do is say the word, and they'll march."

Bolivar shook his head. "Thank you. I know that. But I have no intention of retaining power against the will of the people." With that, he swung his horse in the direction of the street. Head held high, his gaunt face reflecting pride and resignation, the great Bolivar began his ride to exile.

The group in the courtyard fell in behind him. Nobody spoke; the only sound was the clopping of horseshoes on the cobblestones. Even though it was early, many people were already awake and on the streets. They were quiet; some had tears streaming down their faces. The men removed their hats and bowed their heads. A mother, standing in an open doorway, clutched her children to her, as if in fear. Some other children were playing, but when they caught sight of the horsemen approaching and recognized the Liberator, they stopped and stood in silent awe, their game forgotten. A woman who had just returned to her home with an armful of bread was met by her husband, who pointed wordlessly to the entourage silently passing by. The woman whispered, "It's the Liberator! He's leaving us, isn't he? He's going into exile." She burst into tears and fled into their house.

A nun appeared in the doorway of a convent. She took one look at the scene, crossed herself and fell to her knees in prayer. In the distance, a church bell tolled. A woman turned to her father standing beside her in front of their house and gave him a questioning look. "The bell's tolling. Who died, do you know?"

"Yes, my dear—Great Colombia."

On a rise outside of Bogota, Bolivar checked his horse, dismounted and turned to bid farewell to those who had accompanied him. He embraced each one. Most tried in vain to hide their tears, knowing they would never see the Liberator again. Finally, overcome by his own emotions, Bolivar turned and remounted his horse. Silently, he rode away, the ground mist rising around him; the few companions and attendants who were going with him followed behind.

The British Ambassador turned to his aide and said, "He is gone. He is gone. The gentleman of Colombia is gone." He wheeled his horse and, feeling as if he had participated in a dire moment of history, joined other members of the diplomatic corps as they proceeded slowly, funereally back to town.

Chapter 2

Thirty seven years earlier, on a spring afternoon in 1793 Simon Rodriguez caught his first glimpse of the vast plantation house at the heart of the Palacios family's San Mateo ranch in the rich, fertile valley about half way between Caracas and Valencia, in Venezuela. It was a beautiful, low-lying hacienda surrounded by stables, barns, woods and meadows. A creek ran leisurely through the valley, edged by purple flowers.

Perched atop his weary donkey, Rodriguez felt out of place amidst so much beauty: a lone rider in shabby dark clothes and a black, broad-brimmed hat. He was twenty-four—a small man, thin as a broomstick, with long black hair and dark eyes that glinted with mischief, some even said madness. And he was also one of the few people in South America who understood what the American and French Revolutions meant, one of the very few who longed for the time when the colonies of the Spanish Empire would rise as well. Rodriguez was, secretly, a dedicated republican and revolutionary.

As he rode past a fenced-in pasture, he spotted two boys inside the enclosure waving sticks with whittled sharp points at the end. They were taking turns running at a small calf and jabbing the terrified animal. Rodriguez was disgusted and repulsed.

As he moved forward to stop them, another young boy ran into view. He was smaller than the others, dressed in well-worn velvet knee pants, a faded silk shirt and an old woolen vest. Perhaps their younger brother, Rodriguez thought.

"Stop that," the boy screamed in a high-pitched voice. "Stop this minute! Stop that, I say!"

The two bigger boys turned, startled by the shrill cry. Recognizing the newcomer, one boy said, "It's only Simoncito. Let's beat him up."

"No, we'll just get in trouble if we do," said the other.

Coming up to them, the little boy named Simoncito shouted, "You're hurting that calf. Look at him. He's bleeding."

The calf was shaking, his eyes darting about in terror.

"You tied him to the fence-pole," cried Simoncito. "You idiots. What the hell do you think you're doing?"

"We're practicing to be bull fighters, you little creep. Where do you think the great matadors learn their trade?" said the older boy.

Simoncito shouted, "You're a couple of cowards. This little fellow doesn't even have horns. He can't fight back. If you think you're so tough, try pasture number three, where they keep the fighting bulls." He pointed east. "Those bulls have horns. They can gore you to death. Go fight *them.*"

"What do you think we are? Crazy? Those bulls are huge! *You* go fight them."

Without a word, Simoncito strutted toward pasture number three. The other two boys fell in behind him. Rodriguez, hidden by the trees, followed silently until they reached the field. A thick adobe wall enclosed the pasture, leaving no doubt it was the domain of the dangerous fighting bulls, one of which stood idly under a large shade tree. Simoncito took off his vest, snatched a pointed stick from one of the larger boys and clambered over the wall into the field. Rodriguez was too astonished at the boy's audacity to stop him. The other two boys climbed onto the wall and stared in amazement as Simoncito walked toward the large bull.

Finally, the older boy shouted, "Simoncito, stop! We were only fooling!"

The other boy screamed, "Get out of there, Simoncito! That bull will kill you. Please!"

They were scared and frantic now but young Simoncito continued inching toward the bull. Slowly, the boy covered the stick with his vest then waved it menacingly as he had seen the toreadors do. "Toro! Toro!" he shouted. "Ven, toro! Come to me!"

The bull was irritated by the boy's presence and pawed the ground, shuffling backwards in preparation for his murderous charge.

"Stop!" the boys on the fence cried. "We know you're brave. Please get out of there!"

Suddenly a huge black man with a mane of flaming red hair leapt the wall, waving a large, red bandanna and screamed, "Simoncito, get out! Now!"

The bull turned toward the big man, then lowered its head and charged. But the man was young and agile and quickly jumped aside. The bull charged blindly ahead, tossing its head in the air, then turned to charge again, but both Simoncito and the giant were scrambling over the wall to safety. Apparently satisfied, the bull began his leisurely return to the shade tree, swishing his tail in a gesture of victory.

Rodriguez wiped his brow with an old cambric handkerchief. He realized the boy, Simoncito, had retreated not out of fear but out of obedience. He had clearly been taught to obey the enormous black man. Thank God for that, thought Rodriguez. It saved his life. Below, the other two farm boys had run off as soon as the giant appeared. In the road beside the clay wall, the man was shouting at Simoncito, jabbing his finger into the boy's chest, knocking him backwards. "If you ever do a thing like that again, I'll have Mama Hipolita spank you so hard you'll never forget it."

Simoncito looked into the eyes of his tall friend. "Those boys are cruel and cowardly."

"You're right," the man answered. "But you must use your head. Foolishness like that can get you killed."

Rodriguez rode on to the main house, turned his donkey over to a stable boy and was led into the large country manor. He waited in the entrance hall, resting his heavy saddle bags on the floor. They contained mostly books plus the few personal items he would need, including a change of shirts and underwear.

"You must be Rodriguez." The voice from behind startled him.

"Don Carlos?" Simon Rodriguez asked, surprised by the man's appearance. Don Carlos Palacios y Blanco was reputed to be the wealthiest nobleman in Venezuela but, to Rodriguez, he didn't look remotely like an aristocrat. There were no silk or velvet or satin clothes. The heavy-set, muscular man in front of him wore a plain cotton shirt and rough wool trousers tucked into a pair of muddy leather boots.

Don Carlos extended his hand without a smile. He had a ruddy complexion—red faced and stout, probably in his mid-thirties. "You expected someone different," Don Carlos said matter-of-factly. "Well, I'm a worker, Señor Rodriguez. This is a large farm and it can't be supervised from a drawing room."

"Yes, sir. I see your point."

Impatiently Don Carlos called, "Muchacho!" A uniformed manservant instantly appeared from the patio and took Rodriguez's saddle bags.

"Follow me," Don Carlos said curtly, leading Rodriguez through a paneled door at the end of the front hall. "This is my office. Please sit down."

Rodriguez sat in a stiff wooden chair facing Don Carlos's desk. He looked enviously at the wall of books wondering if anyone had ever read them. Before he could ask, Don Carlos said, "I haven't much time so let me explain why I sent for you."

Rodriguez sat up. "Yes, sir. I'm rather curious." In fact, Rodriguez had undertaken a two-day trek for the purpose of finding out.

"I have a nephew, my late sister's son. He's ten. Lost his father when he was three. My sister and my father took care of the boy after that, spoiled him rotten, of course, but they both died last year, so I have the—uh, privilege of caring for the little boy."

"Yes, sir. I understand."

"But, as you can see, I'm far too busy to keep an eye on him and he needs guidance. He's undisciplined and requires constant supervision."

"I see," Rodriguez said uneasily. "His name?"

"His name is Simon, the same as yours. Simon Bolivar."

Rodriguez nodded.

"His full name is Simon Jose Antonio de la Santisima Trinidad Bolivar y Palacios."

"I'll stick to Simon Bolivar, if you don't mind, sir."

"You'll be his teacher. You can call him anything you like."

"Perhaps I should meet the young man," Rodriguez suggested, noncommittally. After all, Don Carlos hadn't even asked him if he would take the job.

"He's supposed to be here by now. God knows where he is. He's uncontrollable."

"You're not worried about him then, sir?"

"No, not a bit," said Don Carlos. "Hipolita, the slave, and Jose Palacios keep an eye on him all the time. Hipolita raised him. He considers her his mother and she treats him like her son. Maybe that's why he's so spoiled. Jose was my late sister's slave. Before she died, she freed him but made him promise he'd look after little Simon and make sure no harm came to him and that's what the damned fool promised. Jose's only twenty but he'll spend the rest of his life looking after Simoncito. Mark my words."

Simon Rodriguez looked puzzled. "Then, sir, Simoncito is well cared for. There doesn't seem to be any need for me."

"You don't understand," said Don Carlos. "They are slaves. They protect Simoncito, as they should. But he requires an education and guidance and, by that, I mean discipline. Lots of discipline."

"Yes sir, but, you know, I operate a school in Caracas. I have 114 students. My assistant is teaching them now."

Don Carlos said, "Most of the year we live in Caracas. I summoned you out here to San Mateo because we won't return for another couple of weeks and I want you to get started now."

"I understand, sir."

To impress the new tutor, Don Carlos added, "Simoncito owns twelve large houses in Caracas, damn it. He also owns more land than anybody in Venezuela. He inherited herds of cattle so big you can't count all the stock, gold mines, copper mines, fields of indigo, sugar plantations, rum distilleries and thousands of slaves to work them. Now do you see why I have to stay up all night going over the damned accounts? It's all for that little . . . nephew of mine."

"You say you and Simoncito will continue to live in Caracas?"

Rodriguez asked. He wanted to make sure he was not being misled.

Don Carlos leaned back in his chair and sighed. "Yes. We stay there most of the year and I travel to the different properties."

"So," Rodriguez began gingerly, "I would stay with you in Caracas and look after the boy?"

Before Don Carlos could answer, the high-pitched yelling of a pre-adolescent boy broke the silence in the hallway outside the study. A second later, the large black man, Jose Palacios, entered the room grasping the arm of the courageous young boy Rodriguez had seen in the bulls' pasture earlier that afternoon. He had changed clothes and now wore clean velvet knee pants with a matching velvet jacket over a white silk shirt.

Palacios nodded to Don Carlos and said, "You told me to bring him right in, sir." With his free arm he touched his forehead in a kind of salute to his patron.

"I certainly did. Thank you, Palacios." Don Carlos nodded, politely dismissing him. Jose, with his wild red hair, left quietly.

At once, Don Carlos, his face flushed with anger, turned to the young Simoncito. "Why weren't you here when I told you to come?"

The boy clenched his fists. "Something more important came up." Then, for the first time, Simoncito noticed Simon Rodriguez. He looked him up and down. His expression was one of complete disapproval.

"This is your new tutor, Señor Rodriguez." Don Carlos gestured to Rodriguez. "He will be in charge of you."

The boy's expression turned to rage. He was still mad at the older boys for wounding the calf. "I don't want a tutor!" he screamed, stamping his foot. "Especially not this scarecrow!" Then, without warning, he kicked Rodriguez in the shin.

"Ouch!" exclaimed the tutor as he grabbed his shin and limped to a chair.

"That does it!" Don Carlos said angrily. He grabbed the boy's two small wrists in one of his powerful hands and unbuckled his leather belt with the other. He paused and looked at Rodriguez who was massaging his shin. "No," he snapped. "You do it. You're his tutor, you discipline him!" He handed his belt to Rodriguez. "Show him who's in charge!"

Rodriguez had not seen any hint of fear on Simoncito's face, not when his uncle had grabbed his wrists nor now, with Rodriguez holding the belt. He realized he respected the boy, even though he was spoiled. But more, he realized how worthwhile it would be to mold him into a man who would stand up to authority without a qualm. What a wonderful revolutionary he would make! Rodriguez handed the belt back to Don Carlos. "I won't be requiring this, sir," he said slowly. "I'll discipline the boy in my own way." He had no intention of disciplining him at all, but he knew Don Carlos Palacios y Blanco would never accept that.

Don Carlos's brow furrowed. He seemed disturbed. "I hope you don't mean you're going to torture the boy. I won't stand for that, you know! Punish him, yes. Sadistic, Inquisition-type rot, no. If you even try anything like that, Jose Palacios will kill you, anyway."

It was Rodriguez's turn to smile. "Don't worry, sir. I'm here to teach this fine young man, not to make him bleed."

Chapter 3

Simon Rodriguez's bedroom was larger and cleaner than any he'd ever slept in. Still, the furnishings were Spartan: a wooden bed with stretched cowhide, a table, chair, small chest, a wash bowl and mirror and a chamber pot. After unpacking his saddle bags, the room still looked empty. To him, it was luxurious.

Rodriguez awoke at seven the next morning and went looking for Simoncito Bolivar y Palacios. By daylight, he could see the house was immense—a rambling, one-story structure built around *bill boyd*

a large square patio that was filled with flowering plants and hedges, crisscrossed by flagstone pathways. Walking to the huge doors he had entered the evening before, he looked out at the fields beyond, stretching as far as he could see.

This is too much for one family, he thought. It simply isn't fair. The Spanish colonial system has to be changed.

Ten-year-old Simoncito was still gloating over his display of courage in the arena when he awoke the next morning. His bed was strewn with yesterday's clothes and the floor covered with wooden toy soldiers engaged in battle.

He had heard people say that the mean boys in the bull-pen were his Uncle Carlos's sons by one of the peasant girls on the farm, but Simon didn't believe it. Uncle Carlos was strong-willed and strict, but those boys were cowards. No Palacios would have behaved as they did. But Simon also realized he could have been killed by the bull. And he knew he'd put his friend, Jose Palacios, in danger, too. If Jose had been killed, Simoncito knew he'd want to die with him. Jose was his only real friend and protector. Young Simon knew he'd done a stupid thing in Jose's eyes, and he regretted it.

A knock at the door interrupted his thoughts.

Standing in the hallway was the tall black-haired man in his old black suit. Simoncito stared warily at his new tutor. "What are you doing here?"

"I'm here to teach you. That is all. I'm here to show you how to savor reading and thinking and how to express your own thoughts."

Simoncito eyed him suspiciously. "You mean you're going to make me memorize a lot of crap and do sums and conjugate Latin verbs?"

Rodriguez shook his head. "Not today." He knew that no matter what he had to put up with, this student could be developed into a thinking person. Perhaps, even into a philosopher who would one day help mold other men's minds.

That afternoon the two Simons walked through the countryside.

"I am going to teach you how to *think,* Simoncito," said Rodriguez. "How to think for yourself. Do you know that in North America and in parts of Europe, men no longer blindly obey the dictates of popes or kings or emperors? They make their own laws, their own philosophies!"

"I think that is treason against the crown," said the boy slowly.

Simoncito's remark had a sobering effect on Rodriguez. "Let me put it another way, then. You are privileged. You don't have to work in order to live. Therefore, you can use your brain to accomplish great things, instead of wasting it on simple tasks."

"What sort of great things?"

"Important things. For your life to be worthwhile, you must accomplish something meaningful. It doesn't make any difference whether it's in art or music or poetry or philosophy, but it must be something outstanding. First, you must learn to think. And to learn to think, you must read the works of eminent philosophers so you can train your mind to reason properly."

Simoncito was confused. He could read and write and conjugate Latin verbs and even construct a Latin sentence. But this stuff his teacher was talking about was way beyond him.

"You can do that, Simoncito. I've seen it in you. With my help, you will learn to love truth, beauty and justice for your fellow men." Rodriguez stopped, realizing he'd gotten way ahead of himself. "You see, Simoncito, the world needs leaders to make change, to right the wrongs of men who came before. Here, we all hate the Spanish Crown. The monarchy is corrupt and decadent. The queen has many lovers, as everybody knows. We have no voice in the governing of our land, and the crown taxes us to death. But nobody here does anything about getting us out from under the oppressive weight of our wretched monarchy. That will require a leader, and we have none."

Simoncito nodded, with a trace of excitement in his eyes.

"You can be a leader, Simoncito. Do you understand what I'm saying?"

"Yes, teacher. I do," said the immature young boy, wanting to please.

He didn't, really. He thought things were just fine as they were.

What if the Spanish Court was corrupt? His family had everything they needed, so young Simon didn't feel he was affected by anything the Spanish Government did. He wondered why Rodriguez wanted native Venezuelans to "govern our land." After all, if somebody else was willing to do it, why not let them.

Chapter 4

In the late 1700's, Caracas, with a population of 35,000, was the third largest city in South America and undoubtedly the most beautiful. Broad, tree-lined streets and large houses, lovely public squares, splendid churches and graceful bridges all gave the city an air of charm. Among the most magnificent homes in colonial Caracas was that belonging to young Simon Bolivar. It faced a large church plaza full of flowering trees, tidy paths and colorful birds singing melodiously. The scene was an endless splash of colors—ruby-red flowers, yellow, blue and scarlet birds, green grass and leaves, orange shrubs and white-washed mansions gleaming in the golden sun. The climate was perfect, pleasantly cool. Bolivar's house sat behind an understated exterior of bare white walls, grilled windows and unadorned doors. Inside were lush interior courtyards, superbly furnished salons and, in the rear, a full stable.

"In these exotic surroundings the next four years raced by quickly but not without incident. Simon Rodriguez decided the surroundings were too opulent for him. He persuaded Don Carlos to enroll Bolivar in his school as a student. The boy was fourteen now and ready"

* * *

Rodriguez was putting the few belongings he had into an old trunk, when his door flung open and in walked Simoncito. "What are you doing, teacher?" he demanded.

"I'm leaving."

The boy stood still. "You can't leave," he said. "You and I are friends. You must stay here. I insist."

"We shall always be friends. You're a good boy, Simoncito. You're mature enough now to attend my school. I will see you every day in my classroom, where you will learn many things and where you'll have other boys your own age to play with."

"It won't be the same. Let's go away. Let's go to San Mateo."

"Let's try it this way first."

"I'm sorry I get angry sometimes and behave badly. I won't ever do it again if you'll stay. I'll be so good, you won't believe it's me."

Rodriguez extended his hand. Bolivar shook it, but with a distrustful expression on his face. Finally, he said quietly, "Teacher, I think I'll like going to your school. Will you still be my best friend?"

"You can count on it."

The little boy embraced his instructor. "Where are you going?"

"I'm going home, of course. To live with my family on the other side of Caracas."

Five days later, 14 year old Simoncito could stand it no longer. He missed Rodriguez terribly and found little comfort in the company of his cold and distant uncle. That night, when the house was quiet, he packed a few clothes, tip-toed to his bedroom window and climbed out. For most of the night he wandered through Caracas in search of Rodriguez's house.

He knew he didn't want to live with his Uncle Carlos who ordered him around and beat him with a belt when he didn't obey. He remembered how kind his grandfather was to him and how his sick mother always kissed him when he visited her in her room. But, then, she would cry. It was only in the slave quarters he had found the freedom and happiness Rodriguez talked about so much. He felt that Mama Hipolita was the mother he would have chosen

13

if he'd had the chance. She laughed a lot and whacked him on his bottom when he was bad. Jose Palacios was like a big brother—and a real friend.

After learning that Simon Bolivar y Palacios had tried to run away on many occasions, the learned judges at the high court of the *Audiencia* ordered him to return to his Uncle Carlos Palacios but Simoncito refused. This was unheard of and threw the court into turmoil. After some discussion, it then decided that the young boy could live with his respected tutor, Simon Rodriguez, where he would receive the attention and education due a gentleman of noble birth. There would be no formal schooling with the other boys. Bolivar would be taught individually—and have the sole attention of his beloved teacher during certain hours of the day. The illustrious gentlemen assumed, naturally, that such a renowned pedagogue as Rodriguez would be living with his wife and family, and perhaps one or two servants, in a normal middle-class home.

Simoncito rubbed his eyes, then sat up on the pile of straw that was his bed in Simon Rodriguez's house. He and the other six children slept in a corner of the main room, crowded with an eating table, five chairs, a makeshift kitchen and a spattering of meager belongings. There were four other tiny rooms, alcoves really, where the older members of the household slept. Though it was still dark inside the house, Simoncito could see the early morning grey through an open window across the room. There were no window panes in the house and he thanked goodness the weather was always temperate. He could smell the rancid odors of frying food and the charcoal smoke coming from the cooking stove at the far end of the room.

He pulled on his trousers and was headed outside to the toilet when a baby awoke with a shrill cry.

"Shut up!" Simoncito shouted, ignoring the fact that his tutor, Rodriguez, was snoring soundly nearby. He noticed Rodriguez's young wife throw off her blanket and rise from the small bed she shared with her husband. Although she managed to keep her back to the rest of the household, she was completely naked until she reached for a flimsy robe hanging from a nail on the wall and put it around her body. She was a lovely girl, not quite fifteen, and for some reason, she left Simoncito short of breath each time he saw her naked. Sometimes, too, she would smile at him, lean over far enough to expose her large breasts and flutter her eyelids at him. This always caused a strange and exciting physical reaction in the 14 year old boy.

"So, you're up, are you?" the voice of Rodriguez's mother-in-law startled him. "I saw you staring at my daughter. You ought to be ashamed! At your age, too. Now, go get breakfast and then start sweeping up."

At the stove, he took a piece of piping hot cornbread. With a broad dull knife, he lathered sugar syrup on top of the bread and ate it hungrily.

"Hey! You took my piece," a high-pitched voice shouted behind him. "That was mine. I put it there." It was one of the three other students who lived with Rodriguez.

Simoncito shrugged and walked away, continuing to devour the bread and syrup. He noted in passing that Lolita, the wife of Rodriguez's brother, had her breast bared and was feeding the baby. In the corner near his own bed, Simoncito saw one of the other young scholars, slightly older than the rest, sleeping under a thin cotton blanket with one of the girl servants. He hoped the boy would wake up before Simon Rodriguez's mother-in-law caught him there.

Simoncito was to sweep each morning but thought the task was boring and stupid. The house was cramped and filthy and impossible to clean. Sweeping did nothing but stir up clouds of dust. Besides, the smells inside the house were becoming intolerable. Rodriguez's mother-in-law insisted on cooking fish heads, tripe and other foul smelling fare. He had to get some fresh air.

Closing the front door behind him, he escaped the shrieking of Rodriguez's two sisters-in-law, fighting over a hairbrush, while Rodriguez's brother, Carlos, shouted, "Shut up! Dammit, shut the hell up!"

Simoncito shivered. He hadn't put on his shirt and the air was chilly. Somebody must have stolen his shirt and his hose, too. It seemed that no one in the house had much to wear, so they took whatever clothes they could find.

The two screaming women came running out the door, one chasing the other. "Give me that! That's mine!"

"It's not. It's mine!" With that, the pursuer grabbed the long black hair of her quarry and yanked so hard the other girl fell over. She jumped up and slapped her tormentor hard. Simoncito clapped his little hands with glee. Both girls turned toward him. On their faces resentment began to dawn. They had no intention of entertaining this bratty little rich kid. Simoncito sensed trouble and took off at a gallop.

Out of harm's way, Simon caught his breath. He smiled.

He liked living with Rodriguez. There was always some excitement going on, especially between the men and girls in the house. He was used to seeing them cuddle under the blankets, and it gave him a strange feeling. He longed to get under the covers with Rodriguez's wife.

Simoncito did not return to the house until ten o'clock that morning when he was due to begin his lessons with Rodriguez. Sometimes the instruction lasted for the rest of the day; sometimes only until lunchtime. Still barefooted and half naked, Simoncito was, nevertheless, on time.

The two walked down the street where they could converse without being overheard by the entire Rodriguez household.

"Tell me, Simoncito, how do you like living in my house, living among the poor?"

"I love it, teacher. Everybody's so busy, nobody has time to pay any attention to me so I can do whatever I want as long as I study my lessons."

Rodriguez was confused. "But your life at home was so luxurious. You had everything!"

Simoncito shook his head. "No. There were always servants overseeing everything I did. I couldn't even go to the privy by myself." Simoncito paused, then looked up at Rodriguez. "Tell me, teacher. You and your family aren't slaves. You're free. So, why do you live like you do? In such a small, foul-smelling house?"

Simon Rodriguez realized Simoncito was still just a little boy.

"Because we are poor," he said. "We do not have enough money to live any other way."

Simoncito walked silently beside his tutor. Finally, his eyes widened as if the sun had suddenly broken through his clouded mind. "This is the injustice you talk to me about all the time," he said. "Until now, I didn't really understand what you meant about the equality of man and all that. Now I do. And I cannot help agreeing with you. Is it because the Spanish rule us?"

"Spain believes in keeping everything the way it is. They violently oppose change in any form. The Inquisition keeps our religion pure, the courts keep our laws obeyed, the viceroy insures that the edicts of the crown are complied with to the letter. As long as we are a Spanish colony, there will be no change. Absolutely no change."

"What is the answer, teacher?"

After a short pause, Rodriguez cautiously ventured, "In the north, the colonies of England revolted and freed themselves. Now, they govern themselves. They have elections, as they call them, and everybody votes for whom-so-ever they please. That means they could vote for Simon Rodriguez and make me their ruler."

"You're talking rot, teacher. Who did the English-Americans elect as their king?"

Rodriguez shook his head forcefully. "No," he said. "No king. They have a president."

"Who is their president, then?"

Rodriguez had to think for a minute. "Washington," he said. "George Washington. He is a very great man."

"Was he a school teacher like you?"

"No. He was a heroic general. He led a Revolutionary Army against the forces of the British Crown and defeated them!"

The two walked in silence for a few minutes, Simoncito obviously thinking hard. "Then, that is what I shall do. I shall become a general and defeat the Spanish and be elected president, just like George Washington. He is now my idol. I want to be just like him."

Rodriguez's remarks about George Washington had fired his

immature, young mind far more than the abstract concepts of liberty and equality, which were far over his head, stored away for another day.

He wondered if Rodriguez was making it all up. If not, he thought, then I want to be just like this George Washington and fight as a general against the crown and be president, whatever that is, and rule the whole of Venezuela. He hoped desperately that it was true!

Upon their return to the Rodriguez house, both Simoncito and Rodriguez were surprised to see four middle-aged men dressed in somber black clothes with white neck bands and sour looks on their faces. "We are from the tribunal, come to inspect the house of Señor Simon Rodriguez. Are you he?"

Rodriguez nodded.

"Then, please sign this paper stating we were here and inspected your house."

After reading the paper, Rodriguez signed it, and the men turned and departed. It was obvious they had already inspected the house and spoken to all of the occupants.

Back before the *Audiencia* once more, Simoncito heard the report read. "The house of Señor Rodriguez is not a large house. In it live Rodriguez and his young wife, two girl servants, Rodriguez's brother and his wife and their baby, another man and his nephew, four boys who study there, Rodriguez's mother-in-law, his brother's mother-in-law and two young sisters-in-law."

Once again, the court ordered Simoncito to return to his Uncle Carlos's house. Simoncito almost cried. For the first time in his life, he'd had a real family. And, although he didn't realize it, much more important to his education and later political thinking, was the fact that Bolivar had actually lived among poor people. He had seen human existence at its earthiest. He had been exposed to the wanton, open sexual alliances between men and women. Living in the Rodriguez household gave him an understanding of human nature that served him well for the rest of his life.

Simoncito pleaded with Rodriguez to return with him to Don Carlos's. "At least we will have peace and tranquility," he argued.

"Not the chaos we had in your house that kept me away from my studies."

"If it's so peaceful and tranquil, then why do you run away all the time?"

Simon was silent for a moment. He took his teacher's hand in his and said, shyly, "Uncle Carlos beats me with a belt whenever I do something he doesn't like."

"Oh," said Rodriguez. "Does he beat you hard?"

The boy nodded. "If he uses an alligator belt, it makes my legs bleed. But if it's leather, it just makes big, red welts. I try not to cry, but sometimes I can't help it." Rodriguez agreed to move in with him.

Chapter 5

In 1797, the air in Caracas was filled with rebellion. Everybody was talking about the French Revolution and the slave revolts in Haiti. A group of well-born Spaniards were sent to Venezuela as prisoners because they'd been accused of republican activities against the degenerate monarchy of Spain.

Wanting to expose his student to revolutionary thought beyond books and philosophers, Simon Rodriguez took young Bolivar to the La Guayra prison to meet these unusual prisoners, most of whom were aristocratic, educated radicals.

Simoncito Bolivar y Palacios had never been in a prison before and was thrilled to see it. "Look how thick the walls are," he whispered to Rodriguez.

The prison was built around a large open square and housed cells, common rooms and dining rooms. Simoncito was amazed at how clean and neat the place was. The prisoners, too, were well-dressed, milling about without supervision, as if they were gentlemen and invited guests. Actually, this was a unique prison, prepared especially for the aristocratic "detainees," who were not actually criminals nor a serious threat to the state. But, because of their radical political ideas, it was considered prudent to confine them comfortably to keep them out of trouble.

In the large lecture hall, men crowded into benches arranged in neat rows. Simoncito was surprised to see faces he recognized as visitors to his uncle's house, though no one acknowledged anyone else's presence. The discourse, delivered by one of the prisoners, was entitled 'The Plain and Simple Principles of the Republican System.' It was far over little Simoncito's head, so he simply laughed and applauded when the others did.

Afterward, as the crowd dispersed, Rodriguez took Simoncito's arm and nudged his way toward a man in the center of a group and called out, "Don Jose Maria. Don Jose."

The man beckoned Rodriguez forward.

"I want you to meet my pupil, Don Simon Bolivar y Palacios," said Rodriguez. Turning to Simoncito, he said, "You have the honor to be presented to Don Jose Maria de España, the gentleman who has organized this important lecture."

Simoncito bowed politely and shook España's hand. Don Jose Maria was a good-looking man in his forties, with grey hair and a trim figure.

"How rewarding to have one so young attend our gathering. We should have more young people hear the lectures so they are exposed to these principals early. The seeds we plant today will grow into towering trees tomorrow, which will bear fruit much sweeter than any we ever dreamed of."

When they were outside the prison gates, Rodriguez peered up at the cloudless blue sky. "There are big things in the air," he said. "Great undertakings, Simoncito."

"If Don Jose Maria de España has anything to do with these great undertakings, I'd stay away."

Rodriguez looked surprised. "Why do you say that?" he asked.

"He's a nice man. But tell me, teacher, what has he done in his long life? Has he written a book? Has he led soldiers in battle? Has he made a fortune by his own efforts? Has he served as mayor? Has . . .?"

"Of course not, Simoncito. He's a gentleman, like you."

"As long as he just talks about those stupid, republican theories, fine. What harm in that? But if he ever decides to execute a plan, don't pay any attention to him. Leading men into revolution takes experience and courage, and I didn't see either of those traits in Don Jose Maria."

"You have learned to think and reason, haven't you?" Rodriguez said. "But, tell me, Simoncito, what makes you think you are a judge of men?"

"I evaluate men by their virtues and not by their clothes or their manners, teacher. I learned that from you."

"And you find Don Jose Maria lacking?"

Simoncito shook his head. "No. He's a nice man, as I said. But I wouldn't hire him to oversee one of my plantations."

Early one morning, back at the elegant Palacios home, Simon Rodriguez woke up Simoncito. "The Spanish republicans have escaped. They're safely out of prison. Now, they plan to raise troops on the outside and help liberate Venezuela."

Simoncito rubbed his eyes and said, "You're crazy. It can't be done. Anyway, you stay out of it. You hear me, teacher?"

"Since when do you give me orders?"

"If you get into this thing and get caught, they'll hang you, and I love you and don't want you hanged. You hear me?"

"I don't think you can stop me, my friend."

"Do you want to wager money on that? I shall tell my Uncle Carlos all about it, and then we'll see what happens."

"He hates the crown as much as I do."

"Hating the crown is one thing. Getting involved in a rebellion is something else, teacher. For God's sake, keep out of any foolishness. Will you promise me? Will you?"

"Yes, my boy. I promise."

There was a loud knock on the door. "Simoncito?"

"It's Uncle Carlos!"

The door opened. "What's going on here?" asked Don Carlos.

It was Simoncito who answered. "Rodriguez just brought me the news. The Spaniards in La Guayra have escaped."

"Is that all? Well, good for them, I say. Let's just hope it doesn't go any further. I'd hate to see Spaniards and Venezuelans butcher each other."

Several weeks later, Simon Rodriguez and Simoncito Bolivar stood in the hall outside their bedrooms, dressed in their nightshirts. "The revolution has broken out," said Rodriguez breathlessly. "I heard shots. I knew it would happen."

"I hope you're not involved."

Rodriguez shook his head. "No. I promised you I wouldn't get into it, and I didn't. If I had, I'd be out there right now."

"Then get back to bed."

Rodriguez nodded. "I think I agree with you now. I wouldn't go outside tonight for anything they could pay me."

"Open the doors in the name of the crown!"

In the Bolivar house, windows flew open and lamps were lit in the early morning greyness. A servant fumbled with the bolts on the front doors. Pulling on his trousers, Don Carlos Palacios hurried toward the front of the house, where he met eight uniformed soldiers. "What in the name of all the demons of hell is this about?" he demanded.

The sergeant in charge removed his cap and bowed differentially. He came armed and with the authority of the crown. But, still, he knew exactly who Don Carlos Palacios was, and didn't want to annoy him.

"We are here to arrest one Simon Rodriguez," the sergeant said. "We understand he lives here and tutors your nephew."

"Arrest him? What the hell for?"

"We have information he is a friend of Don Jose Maria de España, sir."

"So? I know Jose Maria, too. Do you want to arrest me?"

"No, sir. There was a plot. Don Jose Maria was the leader, sir.

We have put down the revolt and captured España and killed or captured most of his confederates. Simon Rodriguez is only a suspect, sir. He was seen at the La Guayra prison with Don Jose Maria. The plot had something to do with the prisoners who escaped, sir."

Don Carlos nodded. "You only want Rodriguez for questioning, then?"

"Yes, sir."

"And if he's not involved, you'll release him?"

"Yes, sir."

Simon Rodriguez appeared. He'd had time to dress but not to shave. His hands trembled slightly. His voice quivered as he said to Don Carlos, "I'll see you later, sir."

"Don't worry."

From the shadows, the sobs of Simoncito could be heard.

"Stop that whimpering and come say *'hasta luego'* to Señor Rodriguez," commanded Don Carlos.

"He already has," interjected Rodriguez. "I ordered him to stay inside."

Don Carlos stood in the door and nodded as the eight soldiers led Simon Rodriguez away. Even though he was one of the richest men in Venezuela, he had no political or military influence. Those matters were the exclusive preserve of the representatives of the Spanish Crown. After the soldiers had left, he muttered to himself, "Sons of bitches!"

The next evening, Simon Rodriguez's pretty fifteen-year-old wife appeared at the Bolivar mansion. She was timid, obviously overwhelmed by the huge residence. Since his Uncle Carlos was away, young Simon, then fourteen, was summoned by the butler, who told him that Señor Rodriguez's wife was at the door, inquiring about her husband, who she understood was in jail. The girl looked truly worried. Young Simon took her hands in his, smiled and said, "Don't worry, Señora. Your husband will be released shortly. The soldiers said so."

"Thank goodness," she breathed, putting her hand to her breast.

"May I take your coat?" asked Bolivar.

The girl seemed reluctant, yet curious. She had never seen such a large house in her life.

Simon removed her coat. Her dress was thin, but it was made of good wool, probably a hand-me-down from some rich parent of one of Rodriguez's pupils. She didn't seem to be wearing anything underneath it, and she had sewn the dress in such a way that it accentuated her breasts and clung to her hips. Bolivar felt his heart beat quicken. "Come," he said, "let me show you some of the house."

The tour ended in young Simon's bedroom. "Have you ever seen such a big bed?" he asked.

She shook her head.

"Want to try it?"

Giggling, Maria Rodriguez threw herself onto the bed, then rolled over onto her back. Her skirt was pulled above her knees. Simon climbed onto the bed and lay beside her, laughing and panting. After that, Maria took charge, and an hour later, the two lay absolutely naked in the bed. Simon had had his first lesson in lovemaking and no longer had to wonder why there was so much wiggling and giggling under the sheets at the Rodriguez house. He was still short of breath.

He could do this all night, he thought blissfully. God, it was wonderful! I think I shall steal Maria away from Rodriguez and do this forever!

He heard the front door slam shut. "Hurry," he whispered. My uncle's returned! I'll show you out the side door. Quick!"

At the door, he asked, "Can I see you alone again?"

She smiled coquettishly. "Perhaps," she said.

Then she was gone.

Chapter 6

Accompanied by Jose Palacios, Simoncito presented himself at the royal prison. It was solidly built and heavily guarded, quite different from the one at La Guayra. He was allowed to enter without difficulty, although the guards eyed Jose with suspicion. Simoncito could practically read their minds: A man that large could raise real havoc before they could bring him under control.

As they walked down a large stone corridor, Simoncito asked Palacios, "Why do you stay with us? You're free, you know. My mother freed you before she died."

"You know I promised your mother I'd look after you."

"Forever?" asked Simoncito.

Palacios grinned. "Until you can look after yourself," he said. "But, you know, little Simon, I shall stay as long as you will keep me."

"It's forever, then," said Simoncito. He reached for Palacios's hand and squeezed it. The big man was overcome with emotion.

The jail was jammed with people, visitors as well as common criminals and prisoners suspected of being involved in the recent uprising. A guard escorted them to the large cell where Simon Rodriguez was still waiting to be questioned. The other men in the cell looked like convicts, Simoncito thought. Some had been mutilated for past offenses; some were missing an ear; a couple wore eye patches, while others kept their hands in their pockets to hide a missing finger or two. The place smelled of human sweat and urine.

Rodriguez was sitting in a corner on the straw but immediately rose to greet his pupil and the giant slave. "This is no place for you," he said sadly.

"It's no place for you, either. What can we bring you?"

"They won't let you bring me anything."

Simoncito whispered, "You weren't involved in this, were you?"

"No. I told you I wasn't. I really wasn't."

"Then you'll be out as soon as they've questioned you."

"That could take weeks. There are so many of us in here, you know."

Two days later, Simon Rodriguez returned to the Bolivar residence, thinner than usual and smelling of prison. Simoncito threw his arms around his teacher and cried, "I can't believe it. You're back!"

"Thank God," Rodriguez sighed with genuine relief. Kneeling on the floor, he looked Simoncito in the eye. "You saved my life, my friend. If you hadn't made me promise to stay out of Jose Maria's plot, I would have joined in. And today, instead of hugging you, I'd be in line for hanging."

"I'm glad I made you promise."

"Me too. Now, Simoncito, I must talk with you seriously."

"What is it, teacher?"

"I want you to continue your reading. Study Voltaire and Rousseau. They have differing views but both have good points. You must sort out your own opinions; you have the intelligence to do it. Forget that Voltaire favored monarchy. Remember he was a liberal thinker, who spent time in jail for expressing his opinions. Read again Rousseau's "Origin of Inequality" and "The Social Contract."

Simoncito nodded. "Man was born free, but he is everywhere in chains."

"Good for you, Simoncito. Continue to read. Continue to think. Digest what you read as you would an excellent dinner. That will give you more strength than any food ever could."

Young Bolivar's face took on a pensive, quizzical expression.

Then, he practically whispered the words, "You're saying goodbye, aren't you?"

Rodriguez nodded. "They let me go," he said slowly. "But that doesn't mean they won't change their minds. I intend to leave Venezuela so the minions of the Spanish Crown can never put me back into that awful jail again, so they can't decide to hang me just for the fun of it. As far as the crown is concerned, I shall disappear."

Simoncito felt as if he'd been kicked in the stomach. He couldn't understand what was happening. "I'll go with you," he said.

"No. You have no worry. You're rich, your uncle is rich, your whole family is rich. If they were to touch you, it would cause more trouble than they'd be able to handle. Me? I'm poor. I've always been poor. And I'm a liberal thinker. They'd hang me, and nobody would raise a finger to stop them."

"But I want to be with you. Please let me come."

Rodriguez stood up. He patted Simoncito on the head. "You've made me very happy. I value your friendship, and we shall meet again. This is not adios. It's *hasta luego*."

"Where will you go?"

"I don't know."

"What about your friend, Jose Maria de España?"

"They will give him a trial and then hang him."

Sitting in the spectators gallery of the court, Simoncito Bolivar y Palacios endured the spectacle of the trial. He was not enjoying it. He thought bitterly that it was his Uncle Carlos who had ordered him to attend as an object lesson which would keep him out of trouble in the future. Uncle Carlos had come to believe that Simon Rodriguez was somehow mixed up in the plot. "Why else would he have run away?" was his answer to every argument and protestation Simoncito put forth. Damn Uncle Carlos. He assumed Rodriguez had filled his nephew's head with a lot of radical claptrap, as he called it, so attending the trial would be most beneficial for the boy.

Simoncito did understand the prosecutor's arguments and realized Don Jose Maria was guilty of treason. He couldn't approve of that. Reverence for the monarchy was too deeply ingrained. Men could curse the government and complain about injustice and preach theoretical republicanism—but to take up arms against the crown? That was unthinkable. All his daydreams of emulating George Washington disappeared in the cold light of reality.

Simoncito leaned forward as the evidence was presented. Apparently, España's group had helped the Spanish prisoners in La Guayra escape with the understanding they would raise a force and return to help España and his confederate, Manuel Gual, revolt against the crown. But when the Spaniards were safely out of custody, they immediately abandoned the Venezuelans and simply faded away. By that time the government had discovered their plans, so the conspirators had to act. The revolt didn't spread very far. It was a plot concocted by intellectuals, badly planned and ill advised, with little popular support. Gual got away, but España was captured. His trial was a mere formality, the outcome inevitable.

At the end of the trial, the King's Magistrate boomed, "For attempted rebellion against the crown, Jose Maria de España is to be dragged by horses to a place of execution, where he will be hanged and his limbs and head hacked off to be displayed on spikes as an example to others."

On May 8, 1799, a strange and unfamiliar entourage emerged from the public prison. First, a group of bewildered friars appeared, their heads bowed in prayer. Next came two columns of armed soldiers and, behind them, a horse, slowly dragging a man wrapped in coarse rags and laid atop two blankets. The procession was joined by the brothers of the Orders of Charity and Sorrows, who carried wine and water, or alms bowls, and cried funereally, "Do good deeds for the man they are about to execute." Two priests spoke alternately to the man in the bundle, who seemed to listen intelligently. Several friars held the corners of the blanket to keep the man in place. A few townspeople, soldiers, children and teachers watched from the square.

Simoncito Bolivar walked beside the procession with the ever-faithful Jose Palacios. Again, his Uncle Carlos had commanded him to go. The bells began to toll. The faces of the people on the street were grave and fearful. The harsh voice of the town crier rose above the rattle of weapons, the chants of the clergy, the tolling of the church bells and the doleful tones of those who prayed for the condemned man's soul. The crier walked ahead of the procession repeating the sentence of the court: "For attempted rebellion against the crown, Jose Maria de España is to be dragged by horses to a place of execution, where he will be hanged and his limbs and head hacked off to be displayed on spikes as an example to others. For attempted rebellion . . ."

Simoncito put his hands over his ears. The entire spectacle revolted him. By now España had reached the foot of the gallows. One of the accompanying priests, España's friend, Jose Antonio Tinedo, helped him to stand. España's hands were tied behind him. His face was white with terror. Surrounded by a circle of officials, the priest bade España cast out pride as he prepared for death. The curate of the cathedral helped him mount the steps. Then, before he died, urged gently by the priest, Jose Maria de España summoned the courage to speak. Simoncito was close enough to hear him say, *"Someday, my ashes will be honored by my country at this very spot. Viva Venezuela!"*

The priest embraced him, covered him with his vestments, then remained with him until the end, until the man convicted of treason had fulfilled the last grisly stipulations of his execution at the hands of the hangman.

Simoncito felt the tears rolling down his face. He felt his throat tighten and his stomach turn as the limbs and head were brought down from the scaffold, wrapped in the same cloth rags that had covered the man on his way to execution.

Simon Bolivar wiped his tears and thought soberly: If this is Spanish justice, I want no part of it. I've seen how bravely this otherwise ordinary man was willing to die for his country. Simon Rodriguez was right. Revolution brings out the noblest in every man.

The simple, plain citizens of Caracas were stunned by what they saw and remained silent and still long after the execution was over. The frightened children huddled around their teachers, watching as those in the procession moved away, silently, sadly.

Chapter 7

In 1801, when Simon Bolivar was seventeen years old, he was sent to Spain to live with his Uncle Esteban. There, he was to finish his education and obtain some Old World "polish." Since his Uncle Esteban was a viscount and active in palace politics, young Simon was exposed to the most decadent, corrupt, degenerate royal court in Europe. It repulsed him so intensely that he developed a life-long hatred of monarchy and everything it stood for.

However, it was in Spain that he was presented to his cousin, Maria Teresa de Toro, the daughter of Bolivar's uncle, the Marques de Toro. At first, Simon was polite but uninterested. Gradually, he realized that, although his cousin was plain, she was able to make herself attractive by combing her raven hair in a bun to accentuate her round eyes. She dressed in colors and styles that enhanced her figure. She was also self-possessed; always calm and comfortable. She sparkled when she spoke. The prettier young girls of the court simply chattered among themselves and giggled a lot, or stared vacuously into space. But Maria Teresa made an effort to be companionable and entertaining, and she was able to discuss the topics of the day intelligently. Before long, she and Simon Bolivar were spending hours together, just talking. When he attempted to seduce her, she rejected his advances firmly but good-naturedly and he never tried again, although before long they were flirting openly until the young Venezuelan nobleman realized he was falling in love. Maria Teresa was the mother he never really knew. She was the unattainable lover he longed for. The pure virgin goddess he now worshiped.

The Marques de Toro closed the door to his study and joined his wife. He was blunt. "Our cousin, Simon, is serious. He wants to marry Maria Teresa."

"Oh! How wonderful!"

The marques smiled. He nodded. "Simon is a little bit short, of course. And he's still a rough diamond from Venezuela, but Maria Teresa can remedy that."

"Darling," said the Marquesa, "Maria Teresa is twenty-one already. She'll die an old maid unless she marries soon. And Simon is so rich!"

The marques smiled again and nodded. "And his lineage is flawless. It's a good match." Then the marques frowned. "But he's younger than Maria Teresa. Will he tire of her later on?"

"Maria Teresa can handle him like a baker handles dough. He's completely enraptured by her. She'll keep him under control for the rest of his life."

Because of his youth—he would not be eighteen until July, and Maria Teresa was three years older than he was—the Marques suggested that Bolivar take a tour of Europe. This he did reluctantly, returning ahead of schedule, ardent to marry his true love. A month later, they wed in one of the most celebrated social functions of the year. They were sublimely happy. Maria Teresa was both wife and mother to Simon. She filled a void in his existence as no one else could have done. They spent all their time together, walking in the parks, horseback-riding, dining alone—all the time laughing and flirting like young sweethearts.

Though they made plans to live in Europe for the rest of their lives, Bolivar had to return to South America in order to claim the inheritance of one of his immense holdings in Venezuela. Neither he nor Maria Teresa would dream of his going alone, so he and his bride embarked together in 1802 on the finest, most elegant vessel in the trade between Spain and her colonies.

They accepted the voyage as an adventure. They met new friends. They walked the decks for hours, whether the winds howled and the seas rose to meet the ship's prow or if they were becalmed and listless in the water. They were happy and even enjoyed the fare served at the captain's table in the ship's dining room. Their world was idyllic.

After disembarking, they stepped into the elegant carriage awaiting them at the dock and began their trip to the Bolivar mansion in Caracas. Passing through the city portals, Simon shivered slightly. "What is it, dearest?" asked Maria Teresa.

"Oh, it's nothing," replied Simon. Actually, he had seen the shriveled head of Don Jose Maria de España, skull exposed, still impaled on a spike. A bad portent, he thought. He had trouble sleeping that night.

Taking up residence in Caracas, their idyll continued. There were balls and dinners in honor of the newlyweds and the usual jokes about when an heir could be expected. Life was bliss. They had been married only eight months and their love was fresh and new. It was the happiest time of Simon Bolivar's entire life.

Then, unexpectedly, Maria Teresa began to complain of chills and fever and terrible headaches. Jaundice set in. She collapsed. Simon was alarmed. He knew the symptoms only too well. Yellow fever. Simon, born and raised in Venezuela, was immune to the disease, but Maria Teresa had spent all her life in Europe and was extremely susceptible. With her hand tightly clasping his, she died. With her last words, "I love you, Simon," their idyll was ended forever.

Disconsolate, Bolivar rose from his beloved's death bed solemnly vowing never to marry again. To friends who called, he gravely intoned, "Heaven gave me an angel without peer. But I was unworthy, so God took her away." To others, he said, "Heaven thought she belonged up there and tore her from me—for she was not meant for this world." As the months passed, Bolivar remained inconsolable, unable to shed the depression and sadness. He withdrew to the huge kitchen of the Caracas mansion where he had spent most of his youth and to Mama Hipolita, the slave who Bolivar considered and treated as his mother. He'd received all the visits of his friends, all the condolences he could bear. Now, he only wanted to be with Hipolita.

One day, Simon sat on a large table in the kitchen. He wore an open white silk shirt and riding britches and boots, indicating he was not going out visiting nor was he expecting visitors. One of his legs hung carelessly to the floor. Hipolita was as black as coaltar but had a lovely face, full lips and a finely chiseled nose. Slim and lithe, she was a good-looking woman. All afternoon she had been trying unsuccessfully to bring her charge out of his silent bereavement. Finally, she said, "Don Simon, for God's sake, say something. Do something. You're dressed to go riding. Go riding."

Looking up dolefully, he said, "The last time I went riding was with Maria Teresa." That ended the conversation.

But Hipolita was shrewd. She and Simon Rodriguez never got along because he tried to seduce her every time they met, and once she had to slap him very hard. But she had to do something to pull her beloved Simoncito out of his now acute depression, and she knew it. "You had a friend here once," she said slowly. "A real idiot. Your tutor. He left, he said, to escape persecution after the de España plot, but I say the reason was because another man caught your Rodriguez in bed with his wife, and Rodriguez had to get out of town fast before the man found him and cut his stupid throat. That's what I think!"

She saw she had Simoncito's attention. She saw him nod his head.

"Yes. Simon Rodriguez. I wonder what ever became of him," said Bolivar.

"I hear he went to Europe. To Paris, somebody told me. You two used to have a lot of fun together as I remember it." She cast a covert glance at Simon without turning her head. Yes, she realized, she was getting to him.

"We had great intellectual discussions. But you're right. It was fun."

This was the first time in several months that Hipolita had seen any sign of enthusiasm from her favorite boy. "Do you think he'll return?" she asked. Hipolita knew he never would, and that Simon understood that, too.

"No. He'll never come back," said Bolivar
"Maybe you could find him in Paris?"

The seed thus planted took root. From then on, the color rose in Simon's cheeks, and his eyes began to shine again. Preparations for such a journey took time, and during that time, Simon Bolivar once again began to take an interest in what was going on around him.

Chapter 8

It was a clear, cold January evening in 1805 in the city of Paris. Madame Fanny Dervieux du Villars, a small, vivacious woman of twenty-eight, was greeting guests at the entrance to her fashionable salon. Absent as usual, was her husband, the baron, who was twice her age and, as a provisioner of Napoleon's armies, was seldom in Paris for Fanny's parties.

The room was already filled with gentlemen, many wearing the uniforms of Napoleon's general staff. A few noble and elegant ladies circulated among them. Eight crystal chandeliers hung from the ceiling and twenty-four shimmering wall sconces bathed the livingroom with a warm, cheerful glow.

Simon Bolivar, now twenty-two, had been introduced to many of the guests, particularly the young beauties who seemed eager to win his favor. Only recently had Bolivar begun to discover that women were irresistibly drawn to him and that he could have any one of them he desired. He had no idea why but accepted the fact with alacrity. Still, he was beginning to tire of the party and would have happily snuck out the back and headed to the gambling tables but he wanted to meet the great naturalist, Baron von Humboldt. Fanny had promised he would be there. Besides, Bolivar didn't want to offend his hostess who had not only invited him into her home but shared her bed with him as well.

Hearing a commotion at the door, Bolivar looked over in time to see the baron enter. Forcing his way through the crowd of admirers Bolivar introduced himself to the famous naturalist, who had just returned from his unprecedented five-year journey in South America. As they shook hands, Bolivar said, "I understand you have just come back from your explorations of our continent. I'm from Venezuela."

The baron smiled. Bolivar's commonplace introduction seemed to confirm von Humboldt's half-formed opinion that the Venezuelan was just another unimaginative social creature. Still, there was something about him that interested von Humboldt. He saw a slender young man who couldn't have been more than five feet five inches tall. He had a high forehead, black hair and dark intelligent eyes. The baron sensed that inside this small-framed man lurked a spark that could be ignited. He said to Bolivar, "I understand you advocate the end of colonialism."

"Yes, sir, I do."

"Where do you get these ideas?" the baron asked.

"Books mostly. Rousseau, Montesquieu, Voltaire. And I learned from my tutor, Simon Rodriguez, that all men are equal and the pretenses we assume are vain attempts . . ."

Von Humboldt cocked his head. It wasn't *what* Bolivar was saying so much as his manner that fascinated him. Bolivar had a kind of magnetism that von Humboldt had seldom encountered.

As Bolivar's enthusiasm grew, his speech quickened. "One of my countrymen once wrote, sarcastically: 'To be educated, it is necessary only to know enough grammar, philosophy, law and theology to write reports, say mass, or wear the dress of a monk.'"

Von Humboldt was laughing now, thoroughly engaged by the young man's energy and zeal. At that moment, several young ladies approached Bolivar, giggling. "Simon, we would like to bring you a drink, or perhaps *hors d'oeuvres*."

Bolivar turned to them and smiled. "May I accept your gracious offer in a little while?" he asked.

After bowing to the women, he turned back to find von Humboldt chuckling discreetly. "Señor Bolivar, you seem to have quite a way with the ladies."

Bolivar shrugged. "God knows why," he said modestly. "It must be because they all know I'm sterile, so they run no risk."

The two men laughed. "Are you?" asked von Humboldt.

Bolivar nodded. "I had mumps and measles, both."

The baron felt someone tugging his sleeve. It was his colleague, Aime Bonpland, who nodded politely at Bolivar then chided his friend. "My dear baron, there are a lot of important people here who are dying to talk to you. Could I possibly tear you away?"

"Yes, yes," the baron said, aware of his obligations. Before leaving, he turned once more to Bolivar. "I agree with many of your views, young man. But people consider me far too liberal. It's the reputation that you, too, will acquire, unless you cool down a bit."

"I can't, sir. It is the way I am."

Von Humboldt smiled knowingly. "I enjoyed our talk. By the way, how much do you really know about your native continent? About South America?"

"I know that as long as it's in the firm grip of Spain, our future is as bleak as our past."

Even as Bonpland was attempting to drag him away, von Humboldt leaned closer and said quietly to Bolivar, "I too believe your country is ready for its independence. The only thing you lack is a leader dynamic enough to achieve it."

That night, as Fanny du Villars lay in bed with Simon Bolivar, she bit his ear tenderly and whispered, "Darling, you've got to learn to be more discreet."

"What do you mean?"

"You mustn't criticize Napoleon in public. He's our emperor."

"He was my idol. But ever since the day of his coronation, I consider him a hypocrite. He betrayed the revolution. A crown is still a crown, nothing but an outmoded custom. Napoleon's fame is a reflection of hell."

Fanny shook her head gently, but Bolivar seemed lost in

thought now, tenderly stroking his lover's naked back. "Just think, Fanny, of the glory that would be won by the man who liberated Venezuela."

Bolivar realized that Fanny was falling asleep. But he could not.

He wanted freedom for his country with all his heart. But, he wondered, was it Venezuela's freedom or his own glory that he desired most. He knew that in Europe he was considered a *colonial*. And he didn't like being a colonial. Unless he could become a famous leader of men, he would live and die an unknown *colonial* farmer.

Chapter 9

A few days later, Fanny du Villars came running into her library, where Simon Bolivar was reading quietly. "I think we have found him," she shouted with excitement. "I think we have found your Simon Rodriguez!"

Bolivar jumped up. "Here in Paris?"

She nodded, still breathless. "And do you know why we couldn't locate the idiot? I'll tell you. Because he calls himself Samuel Robinson now, that's why."

"Samuel Robinson?"

Simon Rodriguez stood in Fanny du Villars's entrance hall, shabbily dressed, his hair uncombed. "Why was I brought here?" he asked the butler.

The man merely bowed and left Rodriguez standing in the hall. He gazed up at the crystal chandelier, then down at the ornate Persian rug and the marble floor. He whistled softly. Then, from the main salon, Simon Bolivar strode toward him, smiling from ear to ear. Rodriguez did not seem to recognize him at first. After all, it had been ten years since they'd parted. Gradually, as he realized it was truly his former pupil, his face split with a large grin. The two men embraced warmly. Afterwards, Bolivar broke their silence. "Who is Samuel Robinson?"

Rodriguez's eyes shifted furtively. "Me," he whispered. "I'm incognito."

"Whatever for?"

"So the agents of the Spanish Crown won't be able to find me, of course. After what happened to Jose Maria España, one can't be too careful."

Bolivar laughed. "Still worried about that, are you? Well, all right, my dear Robinson, let's buy you some decent clothes, then catch up on all these years."

For weeks, they talked incessantly, exchanging stories, experiences, ideas and plans. For Bolivar, it was like breathing fresh air again and he realized how much he had missed his tutor's company and inspiration. At Rodriguez's suggestion, the two traveled to Rome, a city they had often talked about and dreamed of visiting. Their stagecoach journey took over a month.

In Rome, atop the famous Monte Sacro, one of the city's seven hills, the men gazed out over the ruins of ancient forums and theaters, the shells of the old stadiums and slave markets. It was all that remained of the grandeur that was Rome.

Bolivar waved his arm, encompassing the view. "What you see here, Rodriguez, is the work of great men. Men who made and changed history."

Rodriguez nodded gravely.

"Now look at me, teacher. What do you see? You see a rich young man who will spend the rest of his life living off his inherited wealth and accomplishing nothing. Nothing! You see a man who will die and be forgotten as if he had never lived! You see a man from a miserable country ruled by incompetent Spaniards for the benefit of a corrupt Spanish king. You see a colonial farmer who will never be accepted by Europeans. You see a Venezuelan who has no voice in the way his country is governed."

Simon Rodriguez stared at his former pupil and his lips curled into a smile.

"It is time I did something with my life," Bolivar continued.

"And what do you intend?" asked Rodriguez.

Bolivar crossed his arms and said matter-of-factly, "I intend to free Venezuela from Spain." He knew it was an idle boast meant only to impress Rodriguez. He had no plans, nor was he a military leader or a politician. Still, he'd said it, and he realized he was dedicating himself rather casually to a deadly serious ideal. He repented his words almost before they left his mouth.

Rodriguez remained uncharacteristically silent. Finally he asked, "And will this be a good thing for Venezuela as well?"

Realizing his former tutor was taking him seriously, Bolivar answered, "Of course. This is a time for revolution, my friend. Look at George Washington; he freed his country. Look at France; they overthrew their monarchy. Now, it is time for Venezuela."

"And you think you're the man to carry the banner of freedom?"

Bolivar looked around him. The sky was crisp and clear. The city at his feet was rich with history and beauty. He seemed transformed, as if in a reverie. He perceived for the first time that if he was ever to become famous or remembered by history he must dedicate his life to a cause now. At the same time he knew there was only one cause in which he believed strongly enough to undertake. Only one in which he could gain renown.

Before he knew what he was doing, he fell to his knees in front of Rodriguez and declared solemnly, "I swear before you and before God and my native land that I shall never give rest to my arm nor to my soul until I have broken the shackles which chain us to Spain."

His dramatic declaration silenced Rodriguez as he grasped the magnitude of Bolivar's words. At last, he said, "I am proud of you, Don Simon. But remember this: The benefactors of mankind are not born when they first see the light, but rather when they begin to spread it."

Chapter 10

Rodriguez had decided to remain in Italy, and the two men parted, Bolivar going back to Paris. But he missed Rodriguez and was already growing tired of Paris and of Fanny. Bolivar wanted to go home. He was Venezuelan and knew he should take a hand in managing his affairs there.

Fueled by his love for the country and the philosophy of freedom it embraced, he decided to go first to the United States of America, a place which had fascinated him ever since he learned about it as a pupil of Simon Rodriguez. After disembarking in Boston, visiting New York and Philadelphia, Bolivar traveled to Washington, the new capital of the United States. It was March of 1807, and while the town was still small, buildings were going up everywhere, their skeletons silhouetted against the sky. Carpenters' hammers could be heard from all directions, and wagon loads of granite, marble and bricks groaned as they wobbled along the wide roads, covered now with wooden planks to keep the wheels from sinking into the mud.

Bolivar pulled his cloak around his shoulders and shivered in the cold. His guide, a tall, blond American, said through a woolen scarf. "You've been here in the United States three months, you say, Mr. Bolivar? How do you like it?"

"I love it. I'm going to Charleston after this. Every city has been fantastic! In my own country—"

Bolivar stopped as an impressive man with reddish, greying hair, wearing a long brown coat and tan trousers rode by. The guide lowered his scarf, grinned, then waved at the man on the horse.

The man waved back, then nodded pleasantly at Bolivar. As their eyes met, a spark of intellectual recognition ignited between the two men which lasted no more than a fraction of a moment.

"Who's that?" asked Bolivar, genuinely curious.

"Why, that's our president. That's Mr. Thomas Jefferson."

Bolivar's eyes widened. "Without an escort? Without a carriage? Riding a horse alone, like a country lawyer? You must be trying to fool a poor foreigner."

"No, sir! That's President Jefferson, himself."

"The President of the United States? One of the Founding Fathers of the Republic?" Bolivar was stunned. "And he's riding his horse down the street like any other citizen? This truly is equality. This is what the entire world should try to achieve!"

Walking away afterwards, Bolivar seriously considered selling his Venezuelan possessions and immigrating to the United States. It represented everything he wanted. Then, remembering his vow on Monte Sacro, he slowly shook his head.

He knew that such a course of action would be too easy and too selfish. He had to make Venezuela a land of freedom and equality too. He had to keep his vow. But he also knew he'd give anything in the world to live in the United States of America!

In 1808, a year after Simon Bolivar returned to Venezuela, Napoleon put his brother, Joseph, on the throne of Spain. The colonies rejected Napoleon's acquisition of Spain. They began to experiment timidly with self-rule. This presented Simon Bolivar just the opportunity he was seeking. Becoming a statesman, he took up the cause of independence, evolving into a full-blooded revolutionary. In countless meetings across the country, he rarely contained his ambition. Leaping to his feet, Bolivar would exclaim, "You say we must be patient. You say we must wait. Aren't three hundred years long enough? This is our country. We must fight together for its freedom. We are nothing if we are not free." His passion was stirring and contagious and aroused an early show of support from many of the younger Venezuelans.

But the ruling gentlemen of Caracas considered him far too radical. As a form of polite banishment, they appointed him envoy to London in June of 1810. Bolivar accepted the post in hopes of enlisting English aid for the colonies but England had become an ally of Spain in 1809. Therefore, she couldn't support the revolution of one of Spain's colonies.

However, Bolivar did make many influential friends in England, including the Prime Minister and many of the nobility. They lionized him socially. His most important encounter was with a fellow-Venezuelan, General Francisco Miranda, who had fought briefly in the American and French Revolutionary armies and been dismissed by both without distinction. As late as 1806, when Spain was still an ally of France in her war with England, Miranda attempted to invade Venezuela with British backing. He failed miserably, as he always did, and returned to live in England.

Bolivar's letters of instruction from the provincial government specifically forbade him from making any overtures to Miranda. The provisional government in Caracas wanted no part of him. Still the young and impressionable Bolivar was enthralled. He went everywhere with the "famous General Miranda"—the theater, balls, public functions—and the London papers reported the fact. Miranda, in turn, introduced Bolivar to his friends, and the old conspirator and the young revolutionary forged a bond of understanding. The younger man didn't yet comprehend that the older was an adventurer, intent on popularizing himself without regard for any cause or ideal. Against orders, Bolivar returned to Venezuela in 1810 with General Miranda, declaring him, "the man the revolution needs." He was overlooking the fact that Miranda was a man whose entire life had been a series of failures, but who boasted he knew more and had done more than other men.

The political arguments became vehement. Most of the conservative, older members of the councils wanted self-government but under the rule of Spain. Others, more radical and fiery, wanted complete independence. Gradually, the radicals, of which Simon Bolivar was a leader, began to gain the upper hand.

On July 5th, 1811, steered by Bolivar, Venezuela became the first Spanish possession to declare its independence. Simon Bolivar tried mightily to have the declaration signed on July 4th, the same date as that of the United States, but debate and wrangling over minor provisions delayed the signing by one day. General Francisco de Miranda assumed dictatorial powers at the age of sixty-one but proved a weak and ineffectual leader, especially against the Spanish Royalists, whose armies were battling against the revolutionary government with growing success. Gradually, Bolivar realized what a terrible mistake he had made in bringing the incompetent old man back to Venezuela. After months of friction and simmering antagonism, he and Miranda finally came to blows.

It happened on March 26, 1812, when Simon Bolivar stormed into General Francisco Miranda's ornately decorated office. "What the hell do you mean by this?" he shouted, slapping a handbill on Miranda's desk.

The general rose and glared at his aristocratic young subordinate, then picked up the handbill. "It means what it says. We are having a large parade and fireworks. Surely even you can read well enough to understand."

"That's the trouble! I understand only too well! It's going to cost the government more money than we have in the treasury! Look at this office, for God's sake! Velvet curtains, hand-carved desk and chairs. Who do you think you are? The King of England!"

"How dare you?" Miranda roared.

But Bolivar was unyielding. "Last week, you had a formal ball! I don't even want to know what that cost! Balls and parades and other extravaganzas! For what? To promote your popularity, that's what! We desperately need that money to pay the soldiers. We need that money to buy arms. No country will give us credit. How can you be so blind!"

"That's enough! I am in command, and I shall do what I damned well please with what money we have. You take things too damned seriously, Bolivar. You seem to think you're in charge. Well, you're not! I am! Now, get out of my office."

Furious, Bolivar did something he'd never done before. He slammed his fist into Miranda's face. The general staggered backwards, reeling from the blow. Bolivar watched dispassionately, then said, "It's very clear, General, that I should be in charge." With that he turned on his heel and left.

Back in the street, he immediately regretted his impulsiveness. "I never should have hit the old man," he said aloud. "Must be the heat." He looked at the sky. It was clear and brilliant, especially for four o'clock in the afternoon, but the temperature was soaring and the stillness was oppressive. Intending to apologize, Bolivar turned to re-enter the building when he felt a drop of rain on his forehead. He looked up, but there was still not a cloud in the sky. He shook his head and remembered it was Maundy Thursday, the commemoration of the Lord's last supper. He would go to Mass that evening after making confession.

Then, without warning, the earth suddenly trembled beneath his feet. A second later it shook again, this time with such ferocity Bolivar was knocked to the ground. A brick from an adjoining municipal building thudded onto the street less than an foot from his head. Suddenly a tremendous and horrifying roar engulfed the city as houses, churches and public buildings crashed to the ground. People raced out of their homes, screaming and panicked, clogging the streets as buildings continued to sway and tumble.

Bolivar watched in horror as the tall steeple of a small church fell directly onto several dozen fleeing citizens. He heard their desperate screams. Then, as he struggled to his feet, another brick hit him on the back and knocked him flat. Lying on the street, his cheek against the dirt, he was aware of only one thing: the roar and clamor had stopped. There was nothing now but silence.

Suddenly a shrill voice pierced the clear, silent air. "Mercy, King Ferdinand! Have mercy, King Ferdinand! We shall be loyal, King Ferdinand!"

As the moans and cries of the injured began to swell, Bolivar leapt to his feet, galvanized by his love of the republican cause, and shouted, "No! This has nothing to do with the king or anything else! Come help the afflicted!"

Taking off his coat, Bolivar ran to a demolished home and dug feverishly, using his dagger to pry loose the stones and bricks. With his bare hands, he tore away rubble and lifted injured, bloody men and women from the ruins. It was sickening. The first man's face was smashed to a bloody pulp, his arm barely attached to his shoulder. A woman's legs were crushed, her broken bones visible through the ragged flesh. She screamed in agony as he lifted her to safety. From one destroyed house to the next, Bolivar ran, his own hands cut and bleeding, his shirt torn and stained with the blood of those he had saved.

The carnage was unspeakable. Crushed bodies. Screams and groans. Whimpering children covered with blood, their little arms or legs broken, their faces torn.

To those who were too dazed to act, he shouted, "Come! Come help us save lives!" Some were stirred to action but most were too deeply in shock to move. The terror was so intense Bolivar could almost touch it.

In the middle of a street, young Simon came face to face with an arch-royalist priest named Domingo Diaz, who was followed by some thirty adherents. Diaz raised his hands to Heaven and cried, "You see? *The judgment of God!* The judgment of God for defying your king!"

Pushing aside the priest and passing by the crowd surrounding him, Bolivar rejoined, "If nature opposes us, we shall fight it. And we shall force it to our will."

By now the cries of the Spanish loyalists and clergy were more terrifying to Bolivar than the devastation of the earthquake. Everywhere he went, they were calling the disaster, "The will of God." These condemnations of the revolutionary government were not lost on the superstitious populace, many of whom believed that the King of Spain and God were the same being.

"No!" he shouted back again and again. "This is not the will of God. It's an earthquake. They've happened before, and they will happen again, no matter who rules!" Some of the people crowding the streets seemed comforted. They knew Bolivar. They trusted him.

He headed to the main Plaza, walking through streets of utter ruin. So many dead—struck by falling masonry or buried beneath the rubble of their homes.

Nearing the Plaza, he noticed a large crowd paying rapt attention to a monk. Standing atop a bench, the fanatic monk shouted,

"It is for the sin of independence you now pay! You deserted your king and your God to follow these atheistic rebels, these anti-king and anti-Christ republicans!" The crowd nodded raptly.

"No!" shouted Bolivar from the edge of the crowd. "No! We are all Christians! This is not the will of God. It is an earthquake! It is nature!" As he spoke he closed in on the monk.

"That's one of them!" the monk cried, raising his face to the sky. "I call down the wrath of God upon him! God, smite this unbeliever! Strike him down with thy fury! Send lightning bolts to split him in twain." The monk was now shrieking. "Vengeance! Lord have vengeance on this devil! Smite him, Lord. Smite him!"

By then, Bolivar had struggled through the crowd and grabbed the monk. As he pulled him down from the bench, a few onlookers cheered, plainly on the side of the revolution. Bolivar drew his sword. "If I have to kill you to shut you up, you meddling fool, then with the help of God, I will!"

The monk wrenched himself free from Bolivar's grasp and ran into the crowd. Nearby several soldiers were laying out the dead in rows, covering them with blankets, sheets, clothing, or anything else that they could find in the wreckage. Grasping the situation as it unfolded, they interrupted their work and, with gestures and commands, dispersed the bewildered crowd.

Now, standing alone in the plaza, Simon Bolivar smelled the sickly sweet stench of death which shrouded the city. Caracas was flattened. And yet, at this bitter moment, Bolivar knew that the earthquake had destroyed more than the city. It had crushed the revolution. He slumped down on the bench where the monk had stood.

"I will not give up," he vowed. "I will free this land, by God, if it is all that I do."

Chapter 11

Days later, Miranda capitulated to the Royalists and the First Republic of Venezuela was officially overthrown. Miranda fled before he had signed the surrender document. The Venezuelan patriots, including Bolivar, were so outraged by his actions that they turned Miranda over to the Spanish, and he spent the remainder of his life in a prison in Spain. The Spanish reprisals against the patriots were brutal. Before the cruelty of the Spanish manifested itself in their pitiless vengeance against the patriots, the Venezuelans and other South Americans didn't *hate* them. They resented them. They wanted the freedom to govern themselves, but they didn't actually despise their rulers. Now they did! Spanish atrocities made reconciliation impossible. Bolivar fought until he was forced to leave Venezuela.

He fled to Colombia, and from Cartagena issued his famous Cartagena Manifesto, listing the reasons for the First Venezuelan Republic's downfall. Stating that the laws were made to govern a nation of saints, he called for a re-conquest of Venezuela. "A strong government," he contended, "could have changed everything. It would even have been able to master the moral confusion which ensued after the earthquake." He went on, "The forces of revolution should take the offensive. Go, quickly, to avenge the dead, to give life to the dying, relief to the oppressed and freedom to all."

The Cartagena Manifesto gave Simon Bolivar the instant status of a leading political thinker as well as the most dangerous foe of Spain's colonial empire. The Colombians, recognizing Bolivar's hereditary rank of colonel of militia, put him in command of a small force of forty-five men, which he quickly increased by recruiting several hundred patriotic Colombians and Spanish deserters. He had his orders: Under no circumstances was he to wage offensive war against the Spanish, orders he disobeyed immediately.

Although the Colombians had welcomed him and entrusted him with a handful of their soldiers, they were arguing among themselves and engaging in their petty jealousies. They didn't want to complicate things by actually fighting the crown. That was the reason they told Bolivar not to take the offensive but simply hold his line. Hold, hell! thought Bolivar. The Spanish controlled the Magdalena River and unless he could take that advantage away from them they would strangle the patriot army completely. He knew that unless he took offensive measures, he would never free Venezuela. In fact, he might never even see Venezuela again. His determination was to attack, and then attack again, recruiting troops as he moved forward. He realized that the enemy knew his orders as well as he did, so they wouldn't be expecting him. For this reason, he felt he might have a chance to succeed!

After surprising, attacking and taking several small towns, Bolivar's small but tightly knit force, with the help of the town's inhabitants, took Mompox on the Magdalena, assuring the availability of the river to the patriots. Later, he was to say, "I was born in Caracas, but my fame was born at Mompox." He then followed up this success by attacking several more Spanish garrisons and defeating the royalists decisively. With the Magdalena firmly in his hands, he was emboldened to move his meager forces onto the plains facing Venezuela. There his troops lined up, ready to move. One of the Colombian units was under the command of their twenty-four-year-old colonel, a dark-haired, serious minded lawyer named Francisco de Paula Santander, a fervent patriot and already a veteran of several campaigns.

Astride his stallion, Bolivar wrote a note and handed it to his aide. The aide cantered over to Colonel Santander, who was riding up and down the lines of his troops. The colonel's face was slim and angular, his posture stern. The impression he gave was that of a refined, confident leader who had little time for kindness or warmth. The aide touched the rim of his cap and handed the note to Santander, who read it immediately, then looked up and shook his head. But the aide was already on his way back to General Bolivar.

Following him, Santander rode up to Bolivar and saluted curtly. "About this order to march into Venezuela. My men and I fight to liberate Colombia. We won't march into Venezuela."

Bolivar sat still. His horse stretched its neck down to nibble the grass. By then other officers had ridden over to witness the inevitable confrontation. Bolivar looked up, and his eyes locked onto Santander's as he said slowly but clearly, "Either you march, or you had better be prepared to shoot me. Because if you don't, I shall most certainly shoot you."

Santander looked startled. He was a respected Colombian commander. Nobody had ever spoken to him like that. He didn't reply. He pricked his horse's flank and wheeled to return to his troops. He loped unhurriedly until he was in front of the center ranks, then turned to face his men. "Commanders! Prepare to march! Prepare to march!"

Chapter 12

Bolivar was still driving towards Caracas with his army, marching all day, resting at night. In a small farmhouse just inside the Venezuelan border, the evening wind made the flames of the candles on the kitchen table jerk like men being hanged. Seated at the table, surrounded by his officers, Simon Bolivar, recently promoted to brigadier general, displayed an emotion he seldom exposed to the world: rage. He had seen the devastation of his homeland and the mutilated bodies of her dead.

"Those assassins who call themselves our enemies have broken every rule of international law! But the victims will be avenged, the killers exterminated. Our forbearance is exhausted. Our revenge shall equal the cruelties of the Spaniards. Since our oppressors force us into this deadly war, they will vanish from the face of America, and our soil will be cleansed of the monsters who sully it. Our hatred knows no bounds. I declare this to be *A War to the Death.*"

His officers nodded their agreement.

"I have decided to wage this War to the Death to deprive the tyrants of the matchless advantage their system of destruction has given them. They will pay for their crimes in blood!"

Bolivar knew his soldiers would be surprised. He had always been humane, even to his enemies. But it had availed him nothing but defeat, time after time. And he refused to be thought of as a defeated man. He calculated that his *War to the Death* would make the local people think twice before going over to the Spanish and would convince a lot of loyalists to join his cause. Bolivar despised cruelty, but his compassion had not worked. So, he decided to emulate his enemies' harshness to see if that would change his fortunes.

Over the next three months, Simon Bolivar defied all odds and stunned both his loyal followers and ardent enemies. With his initial army of several hundred men growing daily with new recruits, he marched 1,200 kilometers in a country with few serviceable roads, fought six pitched battles against the Spanish forces and won them all. He destroyed five royalist armies, captured fifty cannon, then finally, on August 7, 1813, marched into Caracas at the head of his victorious army. After his *Admirable Campaign,* as it was soon dubbed, Bolivar was the undisputed leader of Venezuela.

Caracas went wild with joy. The cheering crowds, made up of all classes of men, women and children, lined the streets and hailed their new Liberator. As Bolivar passed, an old man raised his silver headed walking stick in jubilation, cupped his hands to the sides of his mouth and shouted, "Long live our Liberator! Long live Venezuela!" The crowd embraced the cry and the chant swelled like a tidal wave engulfing the hero. "Liberator! Liberator! Liberator!"

A group of young women clad in white, with flowers in their arms, rushed up to the young general and seized the bridle of his horse. He gallantly dismounted, accepted the victor's wreath and kissed each of them. One particular girl with a seductive twinkle in her eye caught his fancy and he kissed her twice.

She smiled up at him and said, "You have set us free, General. I'd like to thank you properly if you would permit me." And that very night, she did.

Her name was Josefina Machado, nicknamed Señorita Pepa, and she became a permanent fixture in Bolivar's life for the next several years. Her mother and father were thrilled, assuming that their daughter would marry the eligible young hero. But, like many before her, Señorita Pepa was destined never to become more than a mistress to Bolivar.

Chapter 13

With dictatorial powers, the Liberator was responsible now for governing the country, a position he had coveted for years. But in the meantime, a powerful new threat had formed on the desolate prairie, where the heat melted flesh and the scorpion and tarantula made their home. The men of the plains rebelled against the revolutionary government of Bolivar. In contrast to the republican military novices, the tough, misery-hardened, wretchedly poor plainsmen had been taught to use arms from birth. Knives, spears, lances, rocks—the primitive weapons of the period—were their childhood toys. They were mean, and they were cunning. To survive, they had to fight fiercely and ride hard for days in the glaring sun and pelting rain. When a group of them joined a brigand named Tomás Boves, allied with the Spanish Loyalists, the seared earth became a blazing hell—and soon all of Venezuela was bleeding from their brutality. Suppressing the revolt of the plains became Bolivar's biggest challenge.

Boves infuriated Bolivar. Every time the Liberator heard of Boves's atrocities, his anger and frustration mounted.

Bolivar was horseback riding in a lush valley near Caracas when a uniformed rider of the patriot cavalry appeared on the horizon, then galloped urgently toward Bolivar. The young man was clearly exhausted and terrified by the time he reigned up to the Liberator's horse.

"Sir," he cried, "he has taken Valencia!"

Bolivar felt his face redden with rage. "Tomás Boves again?"

The man nodded. "The devil himself. And his Legion of Hell."

Bolivar sat in his saddle, fists clenched. He seethed. "What I wouldn't give to fight that son of a bitch hand-to-hand with no holds barred!" He knew he was boasting; in a hand-to-hand combat, the butcher, Boves, would slash him to ribbons. No, he would have to shoot Boves down as he would a mad dog.

The city of Valencia was thoroughly looted and pillaged at the hands of Boves and his men. Anybody suspected of harboring patriot sympathies was rounded up for "special" treatment. The younger girls, most still in their teens, were raped, some seven or eight times. The young men were stripped naked and tied to stakes driven into the ground in the central plaza. The girls, after being so brutally ravished, were stripped to the skin and similarly tied to stakes. Boves stood before the rows of wooden posts, laughing as he watched his victims writhe in the agonizing heat. Then, in a loud voice he proclaimed that anyone who tried to set them free would suffer the same fate. They would remain tied in the sun until they died slowly of thirst and exposure. Watching his enemies suffer gave Boves immeasurable pleasure.

Boves's extreme cruelty to the patriots was due to an almost insignificant incident. When Boves originally took his men to join the patriot army, he was rejected by a stupid, arrogant colonel, who not only repudiated him but told him he intended to imprison him. Long ago, in a fit of rage, Boves had apparently beaten his wife to death over her bad cooking, but it had been so many years ago that most people had forgotten about it. Boves rode off with his men and joined the Royalists, who received him gladly and commissioned him a colonel. But he never forgot the sleight of the aristocratic colonel. He swore all patriots would pay dearly for this rejection and insult. His Spanish blood boiled and, as a result, a great many people died horribly.

During the rape of Valencia, a suntanned, blond plainsman rode up to him and said casually, "Goodbye, chief. War's over now. I go home."

"What do you mean?" growled Boves. "The war's not over. We still have plenty of damned aristocratic patriots to kill and torture and rape and . . ."

The plainsman shook his head. "No, chief. I joined your band to get me a shirt. I got it off a rich patriot I killed. Then I got shirts for my brothers. Then one for my father. I ain't got nobody what needs shirts no more. So, for me, the war's over. I'm going home."

Boves laughed. "But how about the girls? You don't like screwing all those girls?"

The man smiled. "There's girls at home, chief."

"And cutting people up?"

"No, that's your game. I like the fighting and the looting and the screwing, but I never did like cutting up people's feet and making them walk on broken glass. Or tying them all to stakes to die in the sun. That's your sport, chief. I go home now."

Watching him ride off, one of Boves's lieutenants said, "You let him go home just like that?"

Boves nodded. "He'll be back—just as soon as his shirt wears out."

One of Boves's cutthroats threaded his way through the stakes until he reached Boves, still enjoying the sight of the youngsters straining against the ropes that bound them to the stakes. "How come we ain't took them big houses yet, chief? Plenty of women in them. Men too. And they're all damned rich gentry, all patriot enemies of the crown."

"Don't worry. I got plans for them," said Boves. "It'll be my little joke on the city of Valencia."

That evening invitations were delivered to all the large houses and mansions of Valencia asking the ladies of the house to a ball in honor of Colonel Tomás Boves. They did not dare refuse. Boves had the power of life or death, and he had proved how cruel he could be. Most of the ladies arrived at the local dance hall with red eyes. When the small band began to play, the music was sad and somber. The hall was dimly lit by candles, creating grotesque shadows throughout the room. The ladies danced with forced smiles; their partners, Boves's "officers," smelled horribly and wore dirty uniform tunics taken from the bodies of slain patriots.

After about an hour, horses's hoof beats sounded in the streets. A few minutes later, there were shots. Then more shots. Fusillades of shots. Inside the hall, the ladies became hysterical. They began shouting to each other, "What's going on?" "Did you hear that?" "What's happening?" Boves's men kept dancing. Now they were smiling.

One matron could take no more. She breathed, "Goodbye. I'm going home."

The others followed her in a stampede.

They returned to empty houses. While they were at the "ball," Boves's men had rounded up all the gentlemen of the town, taken them out to a field and shot them dead.

By the time Bolivar reached Valencia, Boves's army had dissolved. As far as anybody knew, it was nowhere to be found. It had ceased to exist. His tactics were maddening to any enemy, but especially so to Bolivar, who was sworn to catch him and defeat him. But Boves was always too elusive, too cunning. Boves intended to avoid an open battle until he was ready and in a perfect position to win. In the meantime, the regular Royalist Army had returned to the field. They presented a more conventional adversary, and Bolivar was able to bring them to battle just outside of

Caracas. Although outnumbered, Bolivar split the Loyalist armies and defeated the Spanish regulars. But his forces were too depleted to follow up their victory. He was running short of ammunition and supplies. Knowing Bolivar's victory had severely weakened his army, Boves, now smelling blood, struck. At La Puerta, he and Bolivar clashed in a battle between armies of equal strength, about three thousand men each. The battle lasted two and a half hours. Boves ordered his infantry to advance on Bolivar's center and his cavalry to ride against the flanks. Bolivar's army was completely routed. Boves then moved on and took Caracas. Thousands of refugees fled to the east, but, of those who remained, at least four thousand were killed, cruelly tortured to death. Some of the survivors managed to escape to the British West Indies or Haiti, recently freed from French rule. The noblest families of Venezuela left as paupers. Bolivar was among them, fleeing to Jamaica.

Bolivar's uncle, the Marques de Toro, became the head gardener for a rich planter in Barbados. Now in exile, the grandest ladies of Caracas took in washing to earn their living. Many took up sewing or dress-making, while some of the younger ones resorted to a less respectable occupation accumulating more money, clothes and finery than the rest.

Thus it was that the educated class who had instigated the revolution, signed the Declaration of Independence and fought so hard for their ideals, also suffered the heaviest losses of life, money and property. Bolivar's Uncle Carlos Palacios had died of a heart attack when he heard that a band of loyalist insurgents had attacked and sacked his plantation. It was a most sorrowful time. As Bolivar wrote to his Uncle Esteban Palacios, when the latter returned to Venezuela, "You left behind a large and happy family. It has since been cut down by a bloody scythe. You left a country that, newborn, was still nurturing the fruits of its creation, the first elements of an emerging society, and now you find it all in ruins—all a memory. The living have disappeared. The works of man, the houses of God, the very fields of the earth have suffered the terrible havoc. You will ask, 'Where are my parents, my brothers and sisters, my nieces and nephews?' The most fortunate have been buried within the sanctuary of their homes. As for the less fortunate, the fields of Venezuela have been watered with their blood and littered with their bones. Their only crime was the love of justice."

Chapter 14

As soon as the ship tied up to the wharf in Kingston, Jamaica, a tired-looking Simon Bolivar walked down the rude gangplank. He didn't shuffle, and he didn't slink. He was a beaten man, a man perhaps without hope, but if so he didn't show it to the world. On the dock to meet him was his old friend, Maxwell Hislop.

"I heard about your misfortunes." Hislop's tone of voice showed he was deeply sympathetic.

Bolivar nodded and said, simply, "We lost everything but our honor."

He had nothing. In Kingston he shared a room with several other refugees from Venezuela. When invited to dine with the Duke of Manchester, who was Governor of Jamaica, he had to borrow a friend's coat in order to attend. One day Max Hislop dropped by. Without preamble he said, "Simon, you can't go on living like this. I've found you better quarters. A French lady I know has some rooms she rents. Come, let's go take a look at them."

Delighted, Bolivar left with his friend without even hesitating. Inside the lady's house, he surveyed the two spacious rooms. "They're perfect," he declared. "Tomorrow, I shall move my books and the few possessions I have."

"It's starting to rain," said the French lady. "And it looks like an all night downpour. You can sleep here if it doesn't stop."

Back at Bolivar's old room, another friend, Felix Amestoy, arrived looking for Don Simon. "He's not here," one of the roommates told him. "Any message I can give him?"

Amestoy seemed undecided. Finally he replied, "Yes. Tell him I'm afraid one of the ex-slaves who came over here is going to try to kill him. You might know the one I mean. Man named Pio. I think the Spanish have paid him."

"I'll tell him. But, look at that rain! You'd better wait until it's over. By then, Simon might be back and you can tell him yourself."

"I love the rain. I'll just go sit in Simon's hammock on the porch and watch it for a while. If it doesn't stop soon, I'll just have to get wet."

It wasn't long before Amestoy, drowsy from the sound of the cloudbursts, lay down on the hammock and fell sound asleep. He was snoring deeply when a shadowy figure, dripping wet, crossed the porch soundlessly on bare feet. In his hand the man held a sharp dagger about twelve inches long. He stood looking at the hammock for a minute, but the night was far too dark for him to see the face of the sleeper. It took but an instant for the man to thrust the blade into the sleeping man. Amestoy died instantly. He did not even cry out.

The next morning, the ex-slave, Pio, was apprehended and charged with the crime. He confessed, still believing he had assassinated Bolivar and the Spanish would get him out of jail and off the island. When he learned the truth, he showed no fear, and when he mounted the gallows he showed no remorse.

Bolivar, on the other hand, was angry and showed it plainly. However, from that day on, because an attempt on his life had been made and failed, he felt indestructible.

He had finished his famous "Letter from Jamaica" in which he declared, "The bonds that united us to Spain have been severed forever. The hatred which the Iberian Peninsula inspired in us is greater than the ocean that separates us. Unfortunately, Spain has left us a legacy of corruption, inflexibility, vengefulness and greed. But I caution, and this is important, *as long as our fellow citizens do not acquire the talent and virtues which distinguish our brothers to the north, a liberal, democratic system, far from being good for us, will bring us ruin.* I advocate a strong government based on the leadership of strong men. I want unity and liberty, but Unity is more important to me."

The purpose of the letter was to appeal to Europe, especially to England, for assistance and collaboration. As he put it, "And, if we are strong, the world will see that, under the tutelage of some free nation who will help us, we shall develop the virtues that will lead us to glory."

It was a good letter, obviously aimed at the nations of Europe. But the timing was unfortunate. It was 1815, a momentous year, and Europe had more pressing matters on her mind. In 1815 Napoleon returned from Elba to begin his famous one hundred days. The Battle of Waterloo was fought, and the Battle of New Orleans. In Europe and the rest of the world, Bolivar's Jamaica Letter went unheard.

Bolivar moved from Jamaica to Haiti to avoid any more potential Spanish assassins. On January 2, 1816, the day he arrived in Haiti, Simon Bolivar was presented to the President of Haiti, Alexandre Petion. As they shook hands, the physical difference between the two men was striking. Petion was a large, brawny mulatto, the son of a slave, Bolivar a small white aristocrat. Smiling cordially, the Haitian president spoke first. "Welcome to Haiti." Then, grinning even wider, he continued, "The Spanish won't be able to hire anybody here to stick a knife into you."

"You are most gracious, Your Excellency." Bolivar bowed to the president.

"What's this 'Excellency' thing? I thought we were both revolutionaries."

"My revolution failed. Yours succeeded. You won. You are the president."

Petion shook his head. "You are the hope of South America. How can I help you?"

* * *

Three months later, President Petion and Simon Bolivar embraced on the pier. It was clear to the crowd that had gathered that the two men were more than mere allies—they were close friends.

"How can I thank you?" asked Bolivar. "You and you alone have given me what everybody else refused me, money, arms, ammunition, ships, food . . ." Here Bolivar's voice faltered with emotion.

Putting his arm around Bolivar's shoulders, Petion shook his head. "We are best friends. We have much in common, you and I. We both detest the institution of slavery, we both fought for independence."

"You've given me the means to continue to fight for my own country. I shall never forget your generosity to a poor, destitute castaway."

"No, no. You and I are devoted comrades. What I am doing for you, you would do for me were our roles reversed. I am only trying to emulate you."

Again, the two men embraced. Their feeling of fellowship was both sincere and mutual.

Chapter 15

With the arms and ammunition provided by President Petion, Bolivar and his officers were able to gather a sufficient number of refugees to return to Venezuela. There, he lost no time in announcing his presence and recruiting an army. With an audacity bordering on arrogance, he declared that he would hold a Constitutional Convention in the town of Angostura on the Orinoco River and that henceforth Angostura would be the capital of Venezuela. The delegates he gathered around him for the Constitutional Convention promptly elected him president of Venezuela, even though the country was almost entirely in the hands of the Spanish Royalists under their Marshall Morillo.

"Bolivar's back!"

It was General de la Torre reporting to the illustrious Morillo, the renowned hero of the Napoleonic Wars now in command of the Spanish troops sent to subdue the Venezuelan and Colombian uprisings. Pablo Morillo was a short, well-built man with the face and temperament of a bulldog. He was the ideal man for the job of subduing the revolutionary movement—tough, battle-experienced, intelligent, and victorious on the fields of Europe.

His troops were on their way back to Caracas from a campaign which crushed the patriot resistance on Santa Margarita Island. Morillo had subdued Colombia, his subordinates executing hundreds of people in reprisal for their insurrection. Now, inside his tent on the plains of central Venezuela, Morillo leaned back in his camp chair and read the reports by the flickering candles. A look of dismay crossed his face.

Suddenly he rocked forward and pounded the empty ammunition case which served as his desk. "He's back, alright. And he's brought arms and ammunition with him. That Haitian bastard, Petion, *gave* them to him."

"Yes, sir. And now Bolivar and his officers have taken Angostura on the Orinoco," said de la Torre, a tall, thin man with dark, wavy hair and deep-set eyes. Never a brilliant commander, de la Torre was an excellent second-in-command to Morillo's genius.

"Angostura!" Morillo bellowed. "A useless place! They can have it for all I care. Eastern Venezuela's a quagmire. Full of rebel guerrillas. As long as we control the rest of Venezuela—Caracas, the ports, the good farmlands—let Bolivar wither away in Angostura!"

At that moment, an aide entered Marshal Morillo's tent. "Sir, I bring you some strange news."

"Well?"

"General Bolivar has declared Angostura the capital of Venezuela."

Morillo laughed. "We control Venezuela. The capital is still Caracas."

"Yes sir, but General Bolivar has called a congress to meet at Angostura to draft a constitution and elect a legislature."

"What?" exclaimed Morillo. "Such insolence! How can he do that?" Then the marshal paused. He was not a stupid man. He walked a few steps, turned and faced the aide. "Do you realize what he's doing? He's legitimizing his revolution. He's turning a rebellion into a war between two sovereign nations. He's telling Europe and the rest of the world that he intends to rule constitutionally, under law, that he's going to be an elected head of state."

"I believe you're right, sir," said de la Torre, who had been listening and digesting what he had heard. "I'm afraid the world will take notice of what he is doing."

In the new congress hall at Angostura, General Simon Bolivar, who had just been elected president with full powers to rule as he saw fit, was finishing his acceptance speech with the ringing words, "The first day of peace will be the last day of my power!" The crowds were ecstatic; Bolivar was their savior.

Before the masses, he basked in the glory of his success, but in the privacy of his temporary house in Angostura, Bolivar was more guarded. Gathered with his top officers, Rafael Urdaneta and young Colonel Antonio Jose Sucre, Bolivar outlined his aims. "Gentlemen, this will prove to Europe that we're not just a gang of bandits. We have a capital, an elected Congress and a president. We are a nation."

"Do you think the European countries will recognize us?" asked Sucre, a slight but handsome man with poise and modesty.

"If England does, the rest will follow."

"Spain still controls most of Venezuela," said Urdaneta. "Won't the Europeans know that? Won't they just laugh at us?"

Before Bolivar could reply, Antonio Jose Sucre ventured, "I think they know very little of what happens in Venezuela. If they hear that our Liberator has been elected president and we have a Congress, they will naturally believe we exist as a nation."

"Exactly!" exclaimed the Liberator.

After the others departed, Rafael Urdaneta, who was then serving as Bolivar's Military Chief of Staff, stayed behind. He was a tall, slender man with a stern disposition. "General, you don't believe we'll be acknowledged, do you?" Bolivar remained silent. He considered Urdaneta his best military leader after Sucre, whose budding genius was already becoming manifest. He respected Urdaneta's intellect and could always be completely sure of his loyalty.

As a young man, after taking first honors in college, Urdaneta had joined the patriot forces when the independence movement began and fought as a lieutenant in 1811. After the collapse of the revolution in Colombia, he joined Bolivar in 1813 as a major and was promoted to major general the same year. As a youth, Urdaneta had been sensitive and studious and delighted in writing frivolous poetry. But over the past few years he had seen too much war and had forgotten how to smile.

Urdaneta broke the silence. "Sir?"

"Leave me now, Rafael. I must think."

After he'd left, Bolivar sank into his chair and closed his eyes.

He wondered if Europe would even recognize that Venezuela existed. Their rulers knew that Spain still controlled South America and if they decided to throw their full weight against Bolivar's new country, it would be finished. England wanted to help at the beginning, thought Bolivar, but they were allies of Spain against Napolean. Now, they're practical and believe we'll be crushed.

Bolivar slapped his knee. *We need a victory, damn it!* But what chance did they have now that Morillo was in command of the Royalist forces? In Bolivar's mind, Morillo was one of the best military leaders in the world and proved that by beating the armies of Napoleon. Boves is a sadistic, sub-human butcher. But in war, Bolivar realized, he was a bloody genius. Bolivar couldn't even bring him to a full-scale battle. *Still, I am the "Liberator!" Of course I can bring Tomás Boves to battle and whip him.*

But just as quickly, the euphoria faded and the doubts returned.

Bolivar realized he had been concentrating too intensely on his political and military problems. His Congress at Angostura had been a great success. Now it seemed hollow. A mockery. He had to tear his mind away from affairs of state or he would begin to get depressed. "Jose," he called softly.

"Yes, master," came the answer from the darkness outside.

"Stop calling me *master,* Jose. Are there any lonely ladies this evening that you know about?"

"Colonel Amador's widow is still here. She's sitting on her porch, sir."

"Amador died quite a while ago. She must be lonely." Bolivar smiled.

Unbuckling his belt and putting it with his sword and pistol on the bed, the Liberator straightened his tunic, stuck his other pistol in his waistband and strode out of the small house.

Since the Liberator needed practically no sleep, he had remained awake and brooding during the peaceful interludes which followed his lovemaking to Colonel Amador's attractive young widow. He had wanted to return to his room before the sun rose but Clarita Amador wouldn't hear of his leaving before one last amorous coupling which lasted until long after the cocks began to crow.

Several army officers and representatives of the Congress were waiting on his doorstep. "I've been busy," said Bolivar. "Now, I need to think. Please return in an hour."

Removing their hats, the gentlemen bowed deeply before departing. Not one blinked or even smiled, although every one of them knew exactly where their Liberator had spent the night. They backed away, as if they were leaving the presence of a reigning monarch. This always infuriated the Liberator. "Stop that! Put on your hats. We're all equals here."

"Yes, Your Excellency," they murmured, continuing to back away.

Alone, Bolivar gazed out the window on a narrow street. Jose Palacios appeared, ready to offer him breakfast, but when he saw the expression on his master's face he withdrew silently.

I tried to get the United States to send troops to help me, Bolivar thought wearily, but they don't even know we exist.

Suddenly Bolivar was struck with an idea. The English had always liked him and it had been several years since they had beaten Napoleon. The big battle between the French and the English was fought at a place call Waterloo in Belgium. How ironic, thought Bolivar. But that war was over now, so where were all the soldiers? All the arms? Europe had been at war for almost twenty-five years.

Bolivar leapt up from his desk. Europe must be full of out of work soldiers. He could recruit a few regiments of British infantry and knock the hell out of Boves. Then he could tackle Morillo on his own terms with capable, disciplined troops!

Chapter 16

Miles away from Angostura, in the fertile cattle and horse country still controlled by the Spanish Loyalists, lived a twenty-five year old widow named Doña Mercedes. She was the owner, overseer and absolute mistress of the most prosperous horse ranch in Venezuela. Her husband, a staunch royalist, had died early in the war, and since then she had managed their large horse ranch on her own. Doña Mercedes was a competent, head-strong woman with a fierce allegiance to the crown and an extraordinary talent for training horses. Some of her neighbors swore she could school a horse to do anything: lie down and play dead or dance on its hind legs; even defeat the patriots, some said. Dark haired and full bosomed, Doña Mercedes attracted many suitors but kept them all at a distance.

One day, one of her young maids urgently approached her in the training ring. "Señora, we must do something. My husband thinks Boves is in the area."

Doña Mercedes looked across the rambling plains, seemingly

unconcerned. "What makes him think that?" she asked.

"There are riders everywhere. Jose thinks they belong to Boves. He thinks they are getting ready to attack the rebels."

"Nonsense. There are always riders. This is prairie country."

"No, Señora. My Jose, he rode with Boves. He knows."

"So? Boves is a royalist. He's on our side."

The girl shook her head. "Boves isn't on anyone's side. He's out for loot. He's out to kill. Jose quit because Boves's way of killing people made him sick. He wants me to hide."

"Well, hide, then. All of you hide, if that's what you want to do."

Late that afternoon, Tomás Boves and five of his men trotted up the approach to Doña Mercedes's house, a long semi-circular dirt road lined with shade trees and wide enough for three carriages. They dismounted at the door. Since her servants had fled, Doña Mercedes opened the door herself. She bowed to Boves and said, "Welcome, General. This is a friendly house." He was filthy, she noticed, and smelled of sweat, urine and blood.

His voice was gravelly. His breath reeked of garlic and onions. "You're pretty," he said roughly. "I want you." He reached out and grabbed her arm.

"Wait," Doña Mercedes said gently. She was smiling sweetly, but behind the smile, her mind was racing. "I have something better for you. A gift, if you will do me the honor. Please, follow me." She led him to the stables and brought out a magnificent white stallion. "General Boves, I present you with this wonderful war horse. His name is Paco. Consider it a gift from all of us who are loyal to the crown. Please, take him with you."

Boves eyed the horse appreciatively, then led it around to the door of the house. To one of his men, he growled, "Take the horse. Put my saddle and bridle on him in the morning." Then he turned back to Doña Mercedes. "I still want you."

Doña Mercedes did not smile this time. "Well, sir, you can't have me. We may be on the same side in this war, but that gives you no right . . ."

Boves's coarse hand shot out and grabbed the front of her dress and ripped it open, exposing her breasts. Before she could respond, he pushed her through the entrance hall into her livingroom. Doña Mercedes screamed but there was no one to help; they had all retreated in fear before Boves arrived. He threw her to the floor and pulled out a large knife. She was paralyzed with terror as he removed his belt and trousers. There, on the living room floor, he raped her as his men looked on impassively.

Finally, he dragged her up to her bedroom. Several times during the night, her screams pierced the silence, only to be squelched by Boves shouting, "Shut up," followed by the sound of a hard slap.

The next morning, Doña Mercedes, still trembling and in shock, looked out her window and watched Tomás Boves ride off on her white stallion. Though the tears streamed down her cheeks, she gritted her teeth and whispered, "You raped me last night, Boves. But today you will die. And I shall be the instrument of your death."

Later that morning, on a slight rise, Boves sat astride his new white stallion surveying the scene. Next to him was his second in command. His forces were all mounted, positioned nearby and awaiting orders. Below, a column of patriot horsemen rode leisurely along a path. "You know the plan," Boves said to his Lieutenant. "Let them see us. Let them send riders at our flanks to try to hit us on both sides. Then, as soon as they begin their attack, we charge their center. That will finish them." Boves smiled at his own cleverness.

In the distance, the patriot horsemen had seen Boves's men. They halted. Orders were shouted. And, as Boves had planned it, two columns of riders galloped up the rise on each side of Boves's force. The patriot infantry on the road began firing rapidly at Boves's men. At the same time, their cavalry charged up on the flanks. Raising his lance in the air, Boves yelled, "Now!" and urged his horse to lead the charge. His men thundered forward, but Boves's horse didn't budge. Not even a step. He was alone. His men had all plunged forward into the attack. With enemy horsemen closing in on both sides, Boves became desperate. He kicked his spurs into the horse's flanks. He shoved his hands forward. He gripped the horse with his knees. He whipped the stallion's withers cruelly. Still no movement. By now, the enemy was on him. Boves tried to dismount so he could fight on foot, but a patriot lance caught him under his arm and flipped him onto his back. He struggled to get up but another patriot clubbed him on his head and down he went again. As he lay on his back another patriot stood over Boves, raised his lance in both hands and plunged it into Boves's stomach. The metal point dug into the ground, pinning Boves to the earth. By now, Boves's forces were fighting in a wild melee, unaware of their leader's fate. Despite the devastating wound, Boves tried to rise but all he could do was slide his body in excruciating agony up and down the lance pole while the men around him laughed. An hour later, Boves lay alone, dead and stinking in the hot sun.

After Boves had left that morning, Doña Mercedes bathed herself for a full hour, scrubbing her body until it hurt. Now, cleansed and dressed in her best white gown, she stood in her doorway and waved at her workers as they trickled back to the ranch. Most waved back and walked on to the stables or the bunkhouses; others stopped to chat. "Are you all right, Señora?" was the usual question. "Did Boves bother you?"

Her answer was always the same. "No. I was perfectly all right. I gave Boves a horse and he went away."

One man asked, "Which horse, Señora?"

"Paco."

The man smiled knowingly. "You did the right thing, Señora. Boves was a bad man. I don't think he will bother anyone anymore." The man had helped train Paco. Whenever the horse heard even a single gunshot, he would freeze and not move an inch until all was quiet once more.

Much later that afternoon, Doña Mercedes spotted her bloodstained stallion at the end of the dirt carriage-way, his saddle twisted and empty. She smiled and nodded. It was the smile of a person who had exacted just revenge. It was the nod of satisfaction. Doña Mercedes was a formidable woman and no one, she decided, would ever take advantage of her again.

* * *

"I'll be damned," Bolivar sighed. "A woman caused Boves's death!"

"That's what they say, sir."

"Well, however it happened, the revolt of the plains is over."

"Yes, Excellency. Boves's men got their fill of booty and blood and went home. Anyway, Excellency, without him, they were nothing, and they knew it. Boves was the only bond that held them together. Without him they ceased to be a military force."

Yes, thought Bolivar. Boves was the genius who led them to victory after bloody victory. He shook his head. It had been almost impossible to fight against Tomás Boves. After taking and pillaging several towns, his army melted away in twos and threes, riders of the plains, just like any others. Later, they'd all converge and obliterate everything they encountered, then disappear again.

Aloud, he said, "So the butchering son of a whore is dead?"

The officers smiled. In his speeches, letters and proclamations, the Liberator wrote and spoke the most correct and pure language in Latin America. In drawing rooms his conversation was polite and engaging without a hint of vulgarity. But among his soldiers, he could be as profane and uncouth as they. Usually the more ribald his language, the more pleased he was with the way things were going.

"I have to meet this lady," he said. "Please find out who she is and where she is." His speech had returned to its normal, drawing room manner. He wasn't sure whether to be pleased or apprehensive about meeting the woman capable of finishing off Tomás Boves.

Chapter 17

"The devil's coming! The devil's coming! Run for your lives!"

Doña Mercedes ran to the door. "What's this all about?"

"Bolivar, himself, Doña. He's on his way. It looks like he's coming here!"

There's only one road, Doña Mercedes thought, and it leads right to my hacienda. She reflected. Simon Bolivar . . . San Mateo. In the old days we did business with the Palacios family. Simon's mother was a Palacios; and his Uncle Carlos. How bad can he be? I suppose I was young then. He is the leader of the damned revolution, though. War changes people. Maybe he is a devil. Worse than Boves. She shivered and went inside to prepare to flee before he arrived.

The clatter of horses's hooves in the courtyard told her she had tarried too long. Mercedes caught her breath. She crossed herself and dashed to her armoire where she withdrew a stiletto. It was eight inches long with a sharp point. She returned it to its silver scabbard and tucked it under her dress. Mercedes had resolved never to submit again to the kind of abuse she had endured from Tomás Boves. No! She would either kill, if that were possible, or use the dagger on herself if it were not.

When she reached her living room, she heard the men outside knocking on her door. Again, she crossed herself and drew in her breath. "Come in," she breathed softly.

The men did not hear her. They knocked more loudly.

I must face up to this, Mercedes told herself. I'm not as strong as they are, but I'm braver. She shouted, "Come in, I said."

The door swung open slowly, and standing outside in the sunlight were a group of officers in clean, neatly pressed uniforms, wearing their dress sashes and swords. One, smaller than the rest, stepped forward and bowed. His bearing told her he was a gentleman, well-bred and well-behaved. "Doña Mercedes, please allow me to introduce myself, since there are none of your friends here to do the honors. I am Simon Bolivar y Palacios." He bowed and continued, "My only title is that of Liberator, in which I glory."

Mercedes was completely thrown off balance. She looked bewildered, not unlike a soldier ordered to make a suicide charge and then suddenly and without any reason told he didn't have to after all.

Noting her confusion, Bolivar smiled and said, "Mercedes de Garcia y Hoyos, I had hoped you would remember me. I am Simoncito Bolivar. We used to buy horses from your father. Now, I am your Liberator. And since I was in the vicinity, I have come by to pay my respects to you. I apologize for appearing without notice, though I assure you I come as a friend." In actual fact, he had ridden over two hundred miles to see Doña Mercedes.

Relief flooded Mercedes's eyes and a radiant smile appeared on her lips as if by magic. When she spoke, her voice was like music. "Simon, Simon. Of course I remember you. But you're almost ten years older than I am, so you recall the old days better than I do."

Bolivar bowed again. Mercedes, he realized, was lovely. Perhaps the loveliest lady he'd ever seen. He walked slowly into the living room and kissed her hand lightly. "I was so sorry to learn of Pepe's death," he murmured. "How are you faring? Is there anything I can possibly do to be of assistance?" He looked into her eyes and what he saw made him catch his breath. This exquisite creature was attracted to him too. Her mature, voluptuous beauty was something new to him. Her obvious dignity was overwhelming. Without realizing it, he took a step backwards.

"Are you retreating, General? Or are you going to present your officers to me?"

Far from breaking the spell he was under, Mercedes's voice

snatched away the few wits he had left. He stared at her in silent ecstacy. Suddenly he realized Mercedes was moving among his officers smiling and chatting as if they had been her life-long friends. The men were charmed. There was not one who would not have died for her.

Later, after her staff had returned and served Bolivar's party an excellent dinner, Doña Mercedes Machado de Garcia y Hoyos rose from the table. The Liberator's men immediately asked to be excused. They had a lot of work to do, they said, and wondered if they could utilize one of the many unused houses on the farm to work and sleep. This was arranged without delay, as more and more of the hacienda's work force began returning.

After they had left, Bolivar and Mercedes moved to the living room and sat on opposite couches. The room was large. Red and white silk wall coverings gave it warmth and the elegant furniture imported from Europe fit in perfectly. Several beautiful oil paintings were hung between the large windows.

Bolivar lit a cigar. "That smells good," said Mercedes. "I haven't smelled the aroma of a good cigar since Pepe was killed."

"I understand you're still a royalist. Can't you see we Venezuelans should stick together to resist the Spanish?"

"But we are Spanish. You and I were both born Spanish. Besides, I hate revolutions. I like to keep things the way they are. I'm a woman. I need stability in my life. Your damned revolution has cost me my husband and an awful lot more. I hate it, Simon! I hate it. Look what it's done to our country! We were rich and prosperous and happy. Yes! Happy!"

"We weren't free," said Bolivar softly.

"We're not free now. We're all on the point of annihilation. And it doesn't matter whether we're royalists or revolutionaries."

Bolivar realized that, for once, he didn't care if he won the argument or not. All he knew was that he was overwhelmed with desire for this woman. This mature, lovely, graceful woman. He rose from the couch and moved slowly toward her as if by sheer magnetic force. She waited, not breathing, gazing into his eyes longingly. Finally he sat beside her, leaned forward and whispered in her ear, "Mercedes. Mercedes."

She leaned back on a pillow, and sighed. It had been a long time, she thought, since Pepe died. She put her hand up and stroked Simon's cheek. He ran his fingers through her hair. His arms were soon around her in a long, hard embrace. She did not resist. He kissed her full on her lips. Afterwards, he looked into her face and she into his, and perceptibly their eyes began to communicate, expounding feelings no words could express. Mercedes rose and took both of Simon's hands in hers, pulling him to his feet. Together, hand in hand, they ascended the stairs to her bedroom.

The next morning, just before dawn, Simon Bolivar still lay in bed with his arms around Mercedes. He stroked her gently to see if she was awake. She turned and gazed into his eyes. Bolivar kissed her gently and said, "What I wouldn't give to wake up and see you beside me every morning."

Mercedes eyes welled with tears and she looked away shyly.

"I love you," Bolivar whispered.

He hadn't said those words to anybody since his wife had died so many years ago. There had been many women along the way but this was different. Surprised and frightened by his own words, Bolivar leapt out of bed and quickly dressed. As he buttoned his shirt, he looked over at Mercedes, still lying in a sea of silk, and announced, "You're coming with me."

Mercedes sat up in bed with a start. "I want to," she said finally. "I want to more than anything in the world. You and I were made for each other, Simon, darling, but please tell me how I can manage to come with you. I have the horse farm to look after."

"Forget about the horse farm."

"I can't. Too many people depend on me for their living. Let's do this, Simon. You forget about the war. Come live here with me and help me run the farm."

Bolivar buckled on his belt and sword. "Can you sell the farm?"

"To whom? Thanks to you radicals, nobody has any money any more."

"They still consider you a royalist."

"I am still a royalist . . . even though I love you dearly."

Upon hearing the words, Bolivar smiled with genuine happiness. "Why don't you offer to sell the farm to Morillo to keep his army supplied with mounts? He gets tons of money from Spain all the time."

"Are you going to marry me?"

Bolivar looked momentarily confused, then sat on the bed and stroked Mercedes's hair. "I can't," he said at last. "I made a vow after my wife died that I'd never remarry. I won't break it. Besides, you know I can't father children, so why should we marry? I'll promise you this: I shall never leave you and certainly never stop loving you."

"Do you really think Marshal Morillo will buy my farm?"

"Perhaps."

"Then, I'll come to you after I've sold the farm." Mercedes smiled sadly. She loved Simon Bolivar, loved him with all her heart. And she knew he really did love her, too—now. But it wouldn't last. Simon's liaisons never did. No, she'd send him away and promise to come to him. But, of course, she never would. She was too smart to throw away everything she had for a few fleeting months of bliss with Simon. Her evening with him had been exquisitely special. It helped erase the terrible memory of Boves's vicious attack. She would remember the evening spent with Simon for the rest of her life and nothing would spoil it. Still, she knew it was over. Simon would ride away and she would never see him again. A tear rolled down her cheek.

Chapter 18

Bolivar's love affairs were legion. He could not live without a woman, and he didn't care whether she were married or single, as long as she was reasonably good-looking. Except for Mercedes, who gave one excuse after another for staying away, he looked on women, not as companions, but as a means to satisfy his sexual passion. In their embraces, he found relaxation and forgetfulness. Bolivar was all love and charm and grace, completely captivating every woman he met. Yet he was rarely the slave of his amorous adventures. He went from lover to lover as he went from place to place, and he traveled constantly.

Bolivar's favorite mistress at the time was still the lovely Josefina Machado, whom he called affectionately his *Señorita Pepa*. She had become his lover the first evening they met, just after his triumphant entrance into Caracas following the *Admirable Campaign*. Their affair was torrid and passionate.

Early one morning, Señorita Pepa woke up disoriented and unsure where she was. Looking around, she remembered she was in a small tent lying on a blanket beside Simon Bolivar. They had made love for several hours then fallen into the deep sleep from which she now emerged. She smiled to herself. Living with her Simon was glorious. Many were the nights they slept under the stars with no tent over them. Once, she remembered, they had spent almost a week in an old Viceregal palace. After the first night in the palace, both she and Simon had decided they preferred sleeping on a blanket on the marble floor rather than on the ornate and overly soft royal bed. Another time, they'd made love on a ledge behind a waterfall after swimming naked in the splashing water. On the deck of a flatboat on the Orinoco they made love under the night sky. The captain saw them in the moonlight but pretended not to. He turned and strolled away slowly down the deck. Life with Simon was delicious. But she never knew where she'd wake up.

Bolivar and Pepa were always on the move. They went to the West Indies together when they had to flee royalist victories. They shared a hovel together. They returned to Venezuela together. Even so, there were officers in Bolivar's forces who tried to alienate them from each other, claiming Pepa was having affairs with other men, or that when Bolivar went to Jamaica and Haiti, she had resorted to street walking in Trinidad. Those officers were soon relegated to inferior posts. That stopped the rumors. Bolivar and Pepa, together with the patriot army moved constantly, dodging Marshal Morillo and his Chief of Staff, General de la Torre. But, as Simon explained, they'd have to fight soon, or it would be too late for them. Morillo was sure to get more men and arms from Spain.

On March 25, 1818, Simon Bolivar attacked the enemy's Valencia Division at a place called El Semen. The long, costly battle of El Semen pitted Bolivar directly against his rival, Marshal Morillo. For most of the contest, the outcome seemed uncertain as the casualties mounted on both sides. Finally, Morillo gained the upper-hand when his entire reserve army arrived just in time. Though seriously wounded by a lance-thrust in his side, the victorious Marshal Morillo publicly announced that Bolivar's career was ended forever.

A week after the engagement, the patriot troops who survived El Semen were camped a safe distance from their Spanish victors. It was evening. Guards were posted in case of an enemy attack in the night. Camp fires crackled as the soldiers cooked their evening meal. A few tents had been pitched and one- and-two man hammocks strung between trees. Blankets covered the ground where most of the men slept. The talk was subdued, the men sullen.

Next to his tent, Bolivar lay in a wide hammock beside Señorita Pepa while Colonel Francisco de Paula Santander stood talking to them. In contrast to Bolivar's casual attire, Santander wore a stiff uniform jacket and polished boots. Santander still carried the weight of defeat on his shoulders. "After the disaster at El Semen," he said, "I think we're beaten, sir. In fact . . ." he cleared his throat, then continued, "Marshal Morillo has proclaimed the complete victory of the Royalist forces and the end of 'this insignificant little Bolivar's insurrection.'"

Bolivar waved a hand. "We're never beaten unless we think we are. And I don't think we are. You forget, my dear friend, that even the great George Washington was beaten at the beginning of the North American Revolution, not once or twice, but many times. Yet he returned, as I shall return—victorious."

"Forget about your damned George Washington. We lost this battle. We lost our entire infantry, all our war materiel, all your maps and papers and four of your top officers. Sir, with all respect, we are finished."

"You didn't mention that I was almost killed. My aide was blown to pieces riding beside me. If *I'd* been killed, then we would be finished. Venezuela's independence would have been crushed, and with it the hopes of all of South America. But I wasn't. So we shall fight on. And let me tell you something else, my dear Santander. I refuse to be remembered by history as a defeated general."

Santander was seething. "You're always so concerned about what history will say about you! Damn it, sir, we're all about to be routed out and killed."

"No. We shall fight like George Washington did, and we shall win."

Santander shook his head in disgust. "You're not a soldier, sir. You're a ego-driven madman!"

With that, he spun around and stormed away.

Señorita Pepa looked into Bolivar's eyes and asked, "Why do you not dismiss him for speaking to you like that?"

Bolivar smiled. "Because he will come around; he will see that I am right. The man's a lawyer—inflexible as they come. But I need him. He's a fine soldier. Even so, I can't explain my plans to him yet. He won't understand." Then, he lowered his voice in case anybody was eavesdropping outside. "But you do, my dear Pepa. Our experienced British infantry is already on the way to Venezuela."

"I still worry about you, dear," said Pepa.

Bolivar didn't answer. After the failed attempt on his life in Jamaica, and after watching his aide's brains splatter all over him at El Semen, Bolivar felt indestructible. Somehow death always seemed to walk down someone else's path. But he could hardly explain that to Pepa. So he simply hugged her.

Across the camp, Santander heard the commotion as a group of soldiers passed through a sentry post. "Stop," he yelled and the men halted at once.

The sentry approached apprehensively. "They gave the right password, sir."

Santander eyed the soldiers suspiciously. They wore the uniforms of the patriot army but appeared ill-at-ease. "It's after dark. Why are you coming in now? All the survivors of El Semen are already inside."

An officer made his way to the front of the men and tipped his cap to Colonel Santander. "We're looking for the Supreme Commander, sir."

Santander stepped backward. "That's a Spanish term!" Without a moment's hesitation, he turned toward Bolivar's tent and shouted, "General! Run! Treason!"

Shots punctured the night as the Spanish soldiers fired at Bolivar's hammock. Seconds later, one of the men lunged forward, his saber upraised. Santander instantly shot him through the forehead. Bolivar's hammock creaked. A second man ran at Santander. The colonel's sword cut the man down. By now, most of the patriot officers and soldiers in the camp had come to join the fight. The melee was savage, bloody. And short.

Two hours later, General Bolivar and Señorita Pepa rode leisurely back into camp on the horse they had used to escape. Both appeared calm and unruffled and many of the soldiers suspected that they'd used the opportunity for a bit of lovemaking.

After leaving Pepa at their hammock, Bolivar rode on until he found Santander, then swung down off his horse and waited for a report.

"Sir, we got the assassination squad," said Santander. "All seventeen of them."

"Thank you, Colonel," said Bolivar. "I believe you saved my life this evening. I owe you a debt. What can I do to repay you?"

Santander stood straighter and said in a loud clear voice, "Give me my country, sir. I want to lead the invasion into Colombia."

The following evening, Bolivar gave Santander the order to lead his troops into eastern Colombia, take command of the Guerrillas there and fight the Spanish. Santander accepted the commission with deep gratitude and set off with enthusiasm. In the meantime, Bolivar received the information that his English recruits, the finest British veterans of the Napoleonic Wars, were beginning to arrive in Angostura.

He wondered what they'd be like. Not all were veterans; there was a contingent of Irishmen, trained, but too young for Waterloo. They were from good families, Bolivar was told, and wanted to fight for liberty and gain glory. To prove his confidence in them, Bolivar decided to put a few in his Honor Guard. That way, he could get a feel for their abilities and discipline.

Chapter 19

Even before the death of Tomás Boves, a young plainsman named Jose Antonio Paez had gradually taken charge of the fierce horsemen of the prairie. He started out with five men and from that nucleus built a small army, which he led against the

Royalists. Almost unknown to Bolivar and the other patriots, Paez had been operating independently, defeating the Spanish regulars whenever they met. One Spanish officer told of his dismay on facing one of Paez's cavalry charges, describing it as, "a forest of lances spread out at full gallop." By 1817, Paez had made the plains country his private fief.

Riding through the desolate plains, dodging anthills and scorpions' nests, Father Ramon Ignacio Mendez, the future archbishop of Caracas, kept asking the men riding beside him, "Why are you taking me way out here? What do you want from me?"

The answer was always the same. "Uncle Antonio wants you." Even though Paez was only in his mid-twenties, all his men called him Uncle Antonio.

Arriving at a large, flat plain, Father Mendez noted row on row of horsemen arranged in ordered ranks. They were expecting him. There was no doubt of that. Mendez began to tremble slightly, without meaning to. He was thinking: they found out I helped those wounded Spaniards, and they're going to kill me. I'll have to explain I am a man of God and help anybody who's wounded, no matter which side they're on, but my heart is with them. And who in the world is Uncle Antonio? Can they mean the famous Paez? The perspiration began to soak his simple vestments.

A sunburned, blond haired young man rode out to meet Father Mendez. He was smiling as he greeted the priest. As he dismounted, he motioned to Mendez to do the same, then threw his arms about him in a warm welcome. "You know me. I am Jose Antonio Paez, and I have brought you here to receive my oath."

"Your oath?" Mendez sounded perplexed.

"Yes. You see, Father, I have decided to voluntarily submit to General Simon Bolivar as my supreme commander." Paez then waved his hat at the rows of horsemen, who stopped chatting and sat straight, lances held high. He knelt and said, "I, Jose Antonio Paez acknowledge General Simon Bolivar to be my superior and supreme commander. I do this for the good of the country."

Father Mendez improvised by putting his hands on Paez's head and intoning, "I, a priest of the Holy Catholic Church, do bear witness to the oath of our son, Jose Antonio Paez, to put himself under the command of General Simon Bolivar for as long as this war shall last. Amen."

Obviously pleased, Paez yelled to his men, "Now, you chicken chasers, come and do the same. For the good of the country."

To Mendez it seemed like a ceremony reminiscent of a scene from the age of chivalry or the crusades, but he carried on with it as if he did it every day of his life. Afterwards, Uncle Antonio asked the good father to dine with him.

As they sat eating some extremely tough meat, Father Mendez asked, "Do you know General Bolivar well?"

"Never met him," replied Paez.

"Then, why in the name of Heaven have you done this?"

"I did it because I know of his military gifts. The prestige of his name is already known, even abroad."

"This is a gesture, then? Nothing else?"

"Not exactly. You see, Father, to beat the Spanish enemy, we have to unite. We have to fight together. And there is only one man who can unite us—this famous Bolivar."

"I see." But Mendez didn't sound very convinced.

Paez's eyes twinkled. Despite the ceremony which had just taken place, Jose Antonio Paez was never known as a romantic. "You will make certain that General Bolivar knows of this? Won't you?"

Mendez nodded. The task wouldn't be too difficult for him.

"You see, I've heard Bolivar has returned from Haiti with guns and ammunition and food, things that are essential to us, especially since we don't have any right now . . ."

Now Father Mendez understood. He smiled broadly. "I'll see if I can get the general to share these scarce items with his new ally . . ." But Paez was already patting him on the back so hard, it forced the reverend father to cough.

Now twenty-six, General Antonio Paez was the undisputed leader of the plainsmen. There wasn't one of them who wouldn't die for him. Like Boves, he was absolutely fearless in battle, cruel and bloodthirsty as a tiger. The Venezuelan Plains bred fierce warriors. He let his men pillage and loot to their heart's content; having once been as poor as they, he understood their needs.

Paez was known to fight so ferociously that he would become overwhelmed by the intoxication of killing and eventually froth at the mouth, fall off his horse in a glassy-eyed trance, unconscious and ram-rod stiff in what amounted to an epileptic fit. At times like these, his servitor and best friend, a large plainsman named Pedro Camejo, would pick him up and carry him to safety. The two men were inseparable and Pedro Camejo is credited with saving Paez's life more than a hundred times.

When Paez and Bolivar finally met, they admired each other and distrusted each other immediately. Bolivar rode with Paez, fought beside him and continued to admire and distrust him. He admired his fearlessness and physical strength. He distrusted his motives and his loyalty. Paez, for his part, admired Bolivar's brilliance and courage but distrusted him because he was an aristocrat. Therefore, the two men stayed close together, mostly to keep an eye on each other.

In May of 1819, in his hut on the Venezuelan plain, Simon Bolivar sat on a hammock reading dispatches from his officers. On a bench, sharpening a long knife, sat the twenty-six-year-old General Paez, his face burned dark and his blond hair bleached by the scorching sun of the plains. Peasant shrewdness and a native intelligence shone in his eyes. In the corner of the room, Señorita Pepa was instructing several young plainswomen how to care for wounded soldiers.

Pedro Camejo, Paez's closest companion, entered quietly.

"Where have you been, Pedro?" asked Paez. "Out cutting Spaniards' throats again?"

Camejo just smiled.

Pepa shuddered.

Bolivar said, "Pedro, please tell your chief to stop talking like that. It upsets Pepa." He knew Pedro Camejo was the only person who had any influence at all over Paez. Uncle Antonio appreciated him and followed his advice.

"Listen to this," Bolivar said, sitting up straighter and reading from his papers. "Santander says, 'We have defeated the Royalists on the plains of Eastern Colombia. The people have suffered much from the Spanish and are joining us in droves. It is too bad we cannot gain control of the highlands as well but, as you are aware, this is not possible.'"

Bolivar waited for a response from Paez but there was none. "Why is this not possible?" He stood up and walked over to Paez. "Where does Morillo expect to fight us?"

The plainsman shrugged. "On one of the approaches to Caracas, probably Carabobo."

"Suppose we take our army to fight somewhere else? Somewhere he's not expecting us?"

"You've tried to draw part of the Spanish forces south, away from Caracas. You've tried to pull some of their troops to the east. You've been wanting to divide Morillo's army for months now, so you can take on one unit at a time and beat them. But, damn it, you haven't been able to. Morillo's too smart to be deceived by any of your strategies." Pepa looked up thoughtfully. She never liked Paez but veiled her distrust with smiles and small talk. Now she was silent.

Bolivar nodded. "Perhaps. But there is a Spanish army in Colombia, beyond the mountains, and that is the army that maintains Spanish rule in Colombia, right?"

"Sure. Everybody knows that. What are you getting at? The Andes are impassable this time of year. In fact, any time of year."

"Nothing's impossible." Bolivar suddenly seemed infused with determination. "And in the high country beyond the mountains, our attack will be a total surprise. They will be unprepared. You see, they are like you. They believe the Andes are impassable this time of year." Bolivar clapped his hands decisively. "We are going to cross the Andes and defeat the Spanish and free Colombia."

The scarred young plainsman shook his head. "Liberator! You can't be serious. Even I wouldn't attempt such a feat." He never knew what to expect from Bolivar. The man didn't seem to care how many men he lost, as long as he achieved victory—and independence. He wasn't cruel or wasteful, thought Paez. Just single-minded.

Bolivar looked over at Paez and smiled. "Of course I'm serious. Nothing can deter me now!"

Antonio Paez continued to sharpen his knife in silence. He had just been present at the birth of what would become known as one of the most daring feats in military history, and all he could do was shake his head.

Chapter 20

Santander arrived at the rendezvous riding a black stallion. Bolivar was waiting. In the distance, they could see Bolivar's camp—tents dotting the grassy plain of Eastern Colombia. When Santander dismounted, Bolivar embraced him warmly. "Congratulations. You've done such a splendid job here in Colombia, I'm promoting you to the rank of general."

Santander smiled stiffly. "Thank you, sir. And now that we're both generals, I'd like to request the use of some of your forces."

Bolivar frowned. "I've brought my best troops, the British Legion and a thousand men of the plains cavalry, so don't expect me to let you have any of them. I've ordered Paez to send us blankets, boots and wool coats for the crossing."

Santander sighed. "Sir, forgive me for being blunt. But I doubt Paez will send you anything at all. He'll keep whatever he has for himself." He shook his head. "Even thinking of crossing the Andes with this army is ridiculous!"

"I'm doing it. And you're coming with me."

Santander peered up at the clear sky and sighed. "Sir, it's absolute madness. You'll lose most of your army. And probably die yourself."

"Nonsense! My British troops are here. Highly disciplined

men, every one of them. Especially their commander, Colonel James Rooke. Fine man. And that young officer, O'Leary. He's already learning Spanish. Tall, good-looking Irish boy."

"Do they realize what they're in for?"

"Of course not."

The two men rode back. As soon as they arrived, a group of the plains cavalry rode up to Bolivar. They touched their sombreros in a gesture of respect rather than a formal salute and then pointed to the mountains. "We never knew such things existed," said one.

"They're monstrous," said another. "We've never seen anything like them before. We're from the plains."

"They're nothing but masses of rock, and they've got white all over their tops. That's snow. I've heard about it. That means it's terribly cold. And, look at the wind. You can see it blowing."

Bolivar nodded. "Yes, the mountains are high. And they're rugged. They're icy cold on top. And it's going to be extremely difficult to climb them and come out alive on the other side."

"We're glad you agree with us. That's exactly what we're saying."

"But I, Simon Bolivar, am going over those mountains, and I shall fight the Spaniards on the other side."

The men seemed confused. Their leader stepped forward. Colonel Juan Jose Rondon was as black as his horse's mane. Slim, yet powerfully built, the handsome Rondon was renowned for his courage and leadership. Although reluctant to disagree with the great Liberator, Rondon said, "Why, sir? That's all Colombia over there. We're Venezuelans. We don't give a damn about Colombia. Let them fight their own Spaniards."

Bolivar paused. Like everybody else in the army, he respected and admired the fierce Rondon. Finally, he answered his young cavalry leader. "If we beat the Spanish in the highlands of Colombia, we'll be able to free Venezuela. That's the reason Uncle Antonio sent you with me. Do you think he'd have sent you if he didn't think the prize was worth the effort?"

Rondon's finely chiseled face wore a perpetual glare of defiance. Now, he actually smiled and asked, "Do we have any chance at all of making it?"

"I'm going with you. We'd damned well better make it!" answered Bolivar.

Rondon nodded. He knew Simon Bolivar considered himself the life and soul of the revolutionary cause. If he was going over the Andes with them, then he must believe they all had a good chance of making the trek.

After the Liberator had left them, Santander stood with Juan Jose Rondon. The usually stern-faced Santander smiled. "You know, Jose, when Simon Bolivar ordered me into Venezuela at the beginning of the *Admirable Campaign*, I told him the same thing. I said, 'I am Colombian. I fight to liberate Colombia. I won't march into Venezuela.'"

Rondon looked pensive. "Any regrets?"

"Not one. In fact, I'm damned glad I did. If I hadn't, I'd be lying dead in Colombia now with the rest of my friends instead of being a general leading my troops against the Spanish."

Rondon was thoughtful. "What the Liberator was really telling us is this: Neither of our countries can be truly independent without the other."

Santander nodded. He had gained a new appreciation of Juan Jose Rondon's intellect.

Rondon motioned back to the camp and sighed. "Have you seen the shape the foreigners are in?"

"Bad news?"

"The poor men are threadbare—using thorns to hold their uniforms together. My guess is they've never marched through terrain like this before."

"Ah, but they are disciplined! Ragged but disciplined." The two men laughed uneasily.

Just then, two of the foreign officers approached and gave them perfect salutes. Their uniforms were in rags held together by pins and twine. "Sirs," one of them said, "General Bolivar asked us to present ourselves to you." The young man was a fair-haired, red-cheeked boy who spoke Spanish with a strong Irish lilt.

Santander returned a brisk salute. "What's your name, soldier?"

"Lieutenant Daniel O'Leary, sir, from County Cork. I served in the Liberator's Honor Guard at Angostura."

The other officer stepped forward. He was the taller of the two, with wavy dark hair and bright blue eyes. "Sir, Lieutenant Kevin Kelly. Also from County Cork. I *soldier* for a living, sir. No easy Honor Guard life for me."

"How old are you boys?" asked Rondon.

O'Leary answered, "Nineteen, sir. Born in 1800. Makes it easy to keep track."

Santander nodded. "Both of you?"

"Yes, sir," Kelly answered. "Born on the same day."

"In different beds," added O'Leary.

Rondon laughed. "And you've been best friends ever since, right?"

The two boys glanced at each other with the slightest of smiles. Kelly answered briskly, "Depends on the day, sir."

"That's right," said O'Leary. "Some days Kelly's too drunk to even know what day it is."

Kevin Kelly amiably agreed. "After all, we Irish have a reputation to uphold!"

Santander looked thoughtfully at the two young men. Then, pointing to the mountains, he said, "You'll have to look after each other up there, you know. Stick together."

O'Leary asked, "General, what exactly will it be like crossing the mountains of Colombia?"

Santander thought for a moment, then said, "Unimaginable."

"It won't be too hard for Kelly, sir," said O'Leary. "The Kellys cross mountains like that every week just to steal other people's sheep."

"That's because we didn't get rich stealing other people's estates by licking the damned Englishmen's boots like the fawning O'Learys did."

"Fawning, did ye say? Fawning, hell! We got our land by fighting for it, you damned potato farming, sheep stealing . . . "

"Gentlemen," interjected Santander. "This march over the Andes is no joke. It's going to be the hardest thing you've ever done in your lives."

"It can't possibly be worse than the swamps we marched

through to get here," Kelly said. "For a week, we were up to our waists in water. We almost drowned; everybody got the fever; the mosquitoes swarmed over us day and night. Most of the men lost more blood than they did in the Peninsula and Waterloo combined."

Santander and Rondon nodded sympathetically. "It's rough this time of year on the plains," Santander said. "But it's a walk in the park compared to the mountains. You'll need warm clothes, lots of them, or you'll never make it to the other side."

O'Leary shrugged. "I've heard there's none to be had, sir. General Paez was supposed to send us winter uniforms, with blankets and boots and everything else, but for whatever reason, he didn't."

Both Santander and Rondon peered up at the high mountains and shook their heads. Santander turned to his friend. "Maybe Paez thinks if he doesn't send Bolivar the equipment he needs, he won't go."

"Then Paez doesn't know our general very well, does he?"

Santander rode beside Colonel James Rooke, the cheerful, eccentric commander of Bolivar's British Legion. Ahead of them, the plains abruptly ended, cut short by dark masses of rock that soared toward the sky, turning white with snow and ice at the peaks, mingling with the clouds surrounding them.

Colonel Rooke stared in awe. "Biggest damned mountains I've ever seen."

"The Andes," Santander said matter-of-factly. "And we're supposed to go over them. It's June. The Andes are impassable in June. To attempt a crossing is sheer madness."

"You don't want to go?"

"Of course, I want to go. I'm Colombian. But I'm not sure we can go. And I'm not sure if the troops will even be willing to try."

Rooke gazed back at the huge mountains. "Well, if General Bolivar tells me to go, I shall damned well go and take the British Legion with me. I'd follow that man to the end of the world."

"Well, I'll follow him as far as Bogota."

The men laughed.

* * *

Chapter 21

As part of their army, the British had brought a young doctor named Moore. He was a delightful man, good company, and even Señorita Pepa was captivated by him. One evening while the young Irishman, Danny O'Leary, was discussing horses with his friend Kevin Kelly, Dr. Moore and Bolivar sat playing a game of chess. "Check mate!" exclaimed Bolivar. Then he began to cough, as he did more and more frequently.

Moore leaned over so the other men couldn't hear him and said, "That cough, sir . . ."

Bolivar put his handkerchief to his mouth. It came away bloody.

"You know you have tuberculosis, don't you?" asked the doctor.

"I've suspected. My mother died of it. And you already told me how sick my Señorita Pepa is."

"You'll both have to take life a lot easier, sir. Who will you get to lead the expedition across the Andes?"

Bolivar got up and walked around the table. He leaned over and spoke in Dr. Moore's ear. "I shall change nothing. I shall continue to lead my army until I die. And if you say one word of this to anybody, I'll have you shot."

Moore shook his head. "At the rate you're going, I don't think you'll be around long enough to have anybody shot."

"Our little secret," whispered Bolivar. "I don't want the men to find out and lose confidence."

Moore nodded sadly.

* * *

During the early 1800s, tuberculosis was the leading cause of death among all classes of society. Bolivar, like many of his fellow citizens, suffered from the disease. The symptoms were severe: bloody coughing, fatigue, weight loss and fever. Rest and good nutrition were the only remedies. Some of Bolivar's friends attributed his immoderate sex drive to his tuberculosis.

Despite the poor prognosis, Bolivar led an excessively strenuous life, only occasionally succumbing to exhaustion and, even then, recovering quickly. This remarkable stamina under such adverse physical deterioration could only be attributed to the man's overpowering determination and his unyielding dedication to his cause, the same traits which made him such a successful leader of men.

Because she also had the disease, far more critically advanced than Bolivar's, Señorita Pepa had to remain behind when the Liberator departed to cross the Andes. She was growing weaker and weaker, thinner and thinner. Inside an old hut in an Indian village, alone and abandoned, the once vivacious Pepa died. Bolivar's constant companion through his defeats and travails, she perished just before he won his greatest victories.

Chapter 22

The wind and rain lashed mercilessly on Bolivar's soldiers as their columns wound up the most treacherous pass in the Andes, the Pisba. As they climbed, the rain turned to snow and the moisture to ice. Hours into the journey, it became terrifyingly clear that the three thousand men didn't have nearly enough clothes. Icy gales whipped them cruelly, slashing like knives and shredding their uniforms. The narrow trails became slippery and wet. Fog and mist prevented them from seeing anything as they staggered forward. Countless soldiers stumbled to the ground; many never got up. Soon the dead were piled up along the trail, blocking the path of those still able to move; horses slipped off precipices; abandoned supplies and guns lay beside the dying as the snow drifted over them.

Still, the men plodded on, cresting one peak, only to be faced with the awesome and disheartening sight of another still to be scaled. Most soldiers lost all sense of pride and bravery and screamed with pain in the frigid arctic cold.

Santander initially cursed his commander for undertaking such a perilous journey, but before long, he buried his resentment and worked to ease the suffering of his fellow soldiers.

Danny O'Leary and his best friend, Kevin Kelly, never left each other's side. When one started to lag, the other would step in front, clamor up a steep precipice or over the debris left by dead men, then reach back to help the other.

"We'll never make it," Kevin would scream to Danny. "We're all going to die in this God forsaken place. Every one of us!"

"No," Danny yelled back. "We have to keep going!"

The last time Danny shouted his support, a boulder roared down the mountainside and crashed into Kevin, knocking him down a small precipice. Danny raced down the snowy bank to his side. The blood seeping from Kevin's left leg was already frozen; his breath was barely visible, his eyes were fluttering.

"Danny," Kevin whispered, "my leg . . . it's crushed . . . the pain . . . Jesus, Danny, you've got to shoot me."

"No!" Danny screamed. "Hang on! You can make it."

Kelly struggled to grasp his pistol and, with the last of his dying strength, held it out to Danny. "Please, Danny . . . don't make me do it myself."

Quickly Danny reached for the gun, wanting to get rid of it, but the freezing metal had already adhered to Kevin's palm.

Kevin's eyes slowly closed.

"Kevin! Wait! I'll carry you out. You've got to hang on!"

Danny reached down and heaved Kelly's unconscious body

over his shoulder. Slowly, he struggled back up to the trail where the procession of battered, freezing men and horses continued. Danny stepped in line, his body bowed under the weight of Kevin's body draped across his shoulders. Suddenly Kelly regained consciousness and screamed in agony, "Let me die, Danny! For God's sake . . ."

"No!" O'Leary shouted and staggered on. Two soldiers from the British Legion stepped forward to help, but O'Leary insisted on carrying Kevin alone. He kept his eyes on the ground to avoid tripping. Then, through the wind and snow, he heard a familiar voice shouting, "Move, men! Move! Keep moving or you'll die!" Danny O'Leary looked up to see a small man atop a thin but dogged horse, moving back through the ranks. It was Simon Bolivar.

Then, out of the mist appeared the giant Palacios. Pointing to Kelly, he said, "I'll do that, Master O'Leary."

O'Leary shook his head but said nothing. Palacios hesitated then, gently, almost effortlessly, reached out and lifted Kelly from O'Leary's shoulders. Slowly, tenderly, he laid him beside the trail.

O'Leary started to protest. "You can't leave him . . ."

But he was silenced by Palacios's upraised hand. "You've been carrying a dead man, Master O'Leary."

O'Leary sighed, too cold and tired to cry. After a moment, he murmured to Palacios, "Kevin would want us to take his clothes. They'll keep somebody warm."

Danny knew *he* would never be able to wear them.

Reluctantly, he joined the slow parade of soldiers, then, spotting a young Venezuelan plainsman on the brink of collapse, he wrapped Kevin Kelly's coat around him. Next he helped a fellow Englishman into the trousers. Both men looked at O'Leary with silent but profound gratitude which seemed to fill the young man with new strength. He started shouting encouragement up and down the ranks; from time to time, he stopped to wrap a leg, ripped open by the rocky terrain. Through his own display of compassion and strength, Danny O'Leary seemed to be giving strength to the men around him.

Further up the line, a strong and determined Palacios was doing much the same. He removed rocks and boulders that had fallen and blocked the trail. He carried men who wanted to rest, knowing they would never get up if they did. He moved through the columns, encouraging the weak, comforting the dying. When anyone began to tremble and shake, he'd stop and rub them vigorously, restoring circulation.

From time to time Simon Bolivar would appear, praising his former slave respectfully for his efforts. Then he'd disappear into the excruciating wind and snow, shouting, "Keep going, men. Keep moving or you'll die!"

Several times, Palacios had seen Bolivar climb down from his horse and help another man onto the mount. He'd then lift ailing men by their arm pits and virtually drag them forward. Some croaked out in agony, "Let me die. Let me die!" But Bolivar kept encouraging them, lifting them, shouting at them. It was as if the determination of one man could—possibly—get them through the impassable mountains.

Still, the ice and the hard, jagged rocks continued to take a devastating toll. There wasn't a man who was not bleeding; there wasn't a rock or patch of snow without a blood stain. As more men and horses fell, it seemed as if Bolivar's army would perish entirely in the high Andes. Then, again, out of the freezing mist and bitter swirling icy snow, Simon Bolivar would again appear shouting, "Move! Move! You can make it! I know you can make it!" One little man, racked with tuberculosis, was empowering strong, healthy soldiers to persevere and survive.

Bolivar stumbled to his knees as he came out of the mountain pass. He reached down and touched the green grass, as if to make sure it was real. His eyes were red, his skin raw; his uniform was in rags. Behind him, near naked soldiers trickled out of the mountain trails. Bolivar shouted to them, "I'm proud of you, men. Congratulations! You've made it. From here on, it's all downhill."

His words seemed to give the men a new burst of energy. They straightened up; some even smiled through their cruelly chapped lips. An officer emerged. "We lost all our horses."

"I know. I was with you, remember?"

More soldiers straggled down to warmth and safety. Bolivar greeted them as they appeared. "We lost a lot of men," one officer after another told him.

"We had to abandon everything except our guns to cross that last range."

"My men froze to death. They'd never been in weather like that before. They come from the coast."

"General, we were naked. Those winds scourged us like whips. We lost hundreds of men up there."

To him, as to the others, Bolivar said through cracked lips, "I know. I was with you." And, like the rest, the officer nodded. "Yes. You were. We couldn't have made it without you."

The crossing had taken four days of indescribable suffering and agonizing death. A thousand men had died; the dead horses were not counted. But every man who survived knew they had done something momentous.

Chapter 23

Santander limped up to the Liberator's side and pointed back at the massive mountains behind them. After a moment, he said, "There's certainly no going back, is there? We have only two choices now. Either we win—or we die."

The army made its way to a nearby village called Socha. It was primitive and poor, the houses nothing but crude hovels with dirt floors. The smell of wood fires, cooking pots and the sound of crying babies filled the air. The inhabitants hated the Spaniards and welcomed the patriot army as their saviors. They literally gave Bolivar's soldiers the clothes they wore. They sold them their horses. At Bolivar's bidding, they went into the mountains to recover weapons, supplies and ammunition discarded on the long march over the Andes. However, the local Indians' strength was not equal to bringing down the cannon, lodged in crevices and gullies where they had fallen.

After six days in Socha, Bolivar and his army were ready to march. Three thousand men had started the trek over the mountains. Two thousand survived, including six hundred of the newly remounted plains cavalry, led by the intrepid Colonel Juan Jose Rondon. Another six hundred and fifty were Colombian infantry under young General Santander; the rest were British Legionnaires under Colonel Rooke. Santander led the column, followed by the British veterans of the Napoleonic Wars. Bringing up the rear, which was the post of honor, were the plains cavalry. The soldiers hauled the barest of supplies, mostly what they could carry on their backs. A few pack horses were dispersed at intervals among the troops. The pace was steady but slow. Bolivar was now too experienced a commander to exhaust his troops before they even had a chance to fight the battle which lay ahead.

It was July 24, Simon Bolivar's thirty-sixth birthday. A few hours into the journey, a scout returned, pelting down the road toward General Bolivar. Sweat foamed on his horse's flanks and withers. Without dismounting, the man shouted, "General! It's the Spanish army. They're blocking the road at a place called el Pantano de Vargas. Thousands of them."

"Take it easy, son." Bolivar was outwardly calm. "That's why we're here. To fight the Spanish. They've simply saved us the trouble of looking for them."

As the scout rode forward, Santander trotted back from the head of the column to Bolivar's side. He had recovered from the Andes crossing faster than most of the men. Now he sat tall on his horse looking the perfect soldier, his jacket sewn together neatly, his boots shining; even his moustache had been recently trimmed. Bolivar, on the other hand, appeared tired and disheveled, his repressed fatigue and the effects of tuberculosis were visible in every line on his face. Though he never complained, every soldier in his army knew he was sorely debilitated.

"What news?" Santander asked.

"We've run into the Spanish sooner than I expected. They're blocking our route of march and I'm certain they outnumber us. I'm also certain they've picked the best defensive positions in the area. And the commander is probably young Colonel Jose Maria Barreiro. He's good, but inexperienced."

"What shall we do? Their army is between us and Bogota, and the mountains cut off any escape."

Bolivar deliberated for only a moment. "We have no choice but to attack before Barreiro realizes he has such an overwhelming advantage. We have to capitalize on the element of surprise. The Spanish still don't know we're coming or how many we are. It's now or never, my dear Santander."

The dust and smoke of battle swirled around Bolivar. With the crash of each volley, his horse reeled, and bloody, ragged soldiers staggered past him to the rear. Santander rode over and shouted urgently, "The last charge of our infantry failed. We're wiped out, sir."

"No!" Bolivar screamed. "Send me Colonel Rooke."

Rooke cantered up as though he were out for a weekend ride in St. James Park. He touched his cap to Bolivar.

"Colonel, I had hoped to save my best troops. But I have no choice. We've got to break through. The British Legion must attack the heights." Bolivar's face was grave.

Rooke was smiling. "Jolly good, sir." Then he touched his cap again, wheeled his horse and rode off to join his men. When he arrived in front of his troops, he shouted over the din of battle, "Soldiers, today's the day we make a name for ourselves. Fix bayonets!"

The sunlight glistened on the blades of British steel as the Legion moved forward in orderly, unbroken ranks, their commander three paces in front of the first line of infantry. As soon as he saw their files advance, Bolivar swung his horse around and galloped to a position in front of the plains cavalry. He raised his saber and shouted to the gallant black colonel, Juan Jose Rondon, "Save the Fatherland!"

Led by the heroic Rondon, the horsemen who had survived the Andes crossing, six hundred men of the plains, plunged forward in a wild charge, armed only with their long lances. Bareback, they galloped effortlessly through rough terrain. When they reached the Spanish line, the clash of steel echoed through the valley as lance and saber met in a desperate, murderous melee. Finally, after hundreds of men were mortally stabbed, it was the Spanish cavalry who left the field.

At the same time, the Spanish infantry on the heights began to withdraw under the onslaught of the British Legion's bayonets. Watching the Englishmen turn the tide of battle, Santander breathed, "Magnificent! Magnificent!"

When he saw the road was cleared of Spaniards, and their access to Bogota wide open, Santander returned to Bolivar. He leaned over and, in a rare display of sentiment, embraced the Liberator. "We have won," Santander pronounced.

"No. Nobody won. Nobody lost. The road is open, but the Spanish Army got away. We were not strong enough to stop them." Bolivar paused. "But I'll tell you one thing. The Spanish are shaken. They know they can lose now. And our men know they can win. No matter how staggering the odds, our men now know they can beat the enemy!"

Night had fallen. In an open pasture near the battlefield of Vargas, torches burned; campfires blazed. Indian women moved among the wounded, comforting them until the doctors had time to dress their wounds. Burial parties were scouring the field for patriot dead, their torches bobbing in the distance. Daniel O'Leary, his chest bandaged from a saber slash, made his way through the camp to Bolivar's tent and command post. The General stepped outside just as O'Leary arrived. The young man tried to salute but winced in pain and abandoned the formality.

"What is it?" Bolivar asked sharply, clearly irritated at the intrusion. His face was lined with exhaustion.

"Sorry, sir. While I was getting my wounds dressed, I saw Colonel Rooke, sir. I thought you should know he got hit pretty bad."

"How badly?" asked Bolivar. His tone was softer now.

"Shattered arm, sir. Doctor Moore had to take it off."

"How is he doing?" Without waiting for an answer, Bolivar took O'Leary's arm and walked towards the area reserved for the British wounded.

"He's doing fine, sir, but I think he's drunk. Just a little bit, you understand, from the brandy the doctor gave him to take off the arm."

Bolivar smiled. But when they arrived, Rooke was already asleep on his blanket.

Three days later, he was dead.

On August 7th, 1819, Bolivar and Santander dashed ahead of their troops, racing against the Spanish for the bridge at Boyacá. Now, while their horses panted, the two men sat in their saddles, calmly watching the Spanish Army approach the defile before the bridge.

Santander turned to Bolivar and said, "You've played the Spanish like a great toreador. Now, it's you who block them from returning to their base in Bogota." As he spoke, the enemy began to cross the bridge. "Should we move now, sir?"

"Not yet," Bolivar answered quietly. "I want several of their units to cross before we hit them. That way they'll be divided, half their forces on this side, the other half on the far side, and we'll strike before they can reunite."

The two men watched the Spanish advance until several battalions had crossed the bridge. "Now I think we can proceed," Bolivar said calmly. "Jose Rondon will charge against the Spanish on the other side of the bridge, and they've never been able to defeat him yet."

"With your permission, sir, I shall lead the attack on this side."

Bolivar nodded approvingly. "God speed."

Riding to the front of his troops, General Santander raised his saber. "Charge!" he yelled, and immediately galloped toward the Spanish line, wielding his sword to urge his men on. His horse's gait gathered momentum until he reached the enemy. Then, as he reached the front lines, Santander's sabre flashed down again and again, slaying countless soldiers until he fought his way well into the Spanish ranks. His men followed his example.

Fighting fearlessly, they broke the Spanish line. They crushed the Spanish Army.

Covered with blood and grime, his uniform shredded, General Santander rode slowly back to Simon Bolivar, his weary horse desperate for a rest.

Dismounting, Bolivar embraced Santander. "Your courage carried the day. For a lawyer, you fight damned well."

"Our victory here means everything to me. I'd have crossed the Andes ten more times to win today."

Chapter 24

On August 10, 1819, Simon Bolivar entered the city of Santa Fe de Bogota as its liberator and hero. Founded in 1538, Bogota was still a small city of twenty thousand people, built around a large central plaza on which stood the Viceregal Palace, the Cathedral, the Archbishop's residence and other important buildings. Its streets were lined with monasteries, churches, and one-story, whitewashed houses, solidly constructed against earthquake and the mountain cold.

As the victorious Bolivar rode toward the central plaza, followed closely by General Santander, crowds stood in the narrow, cobblestoned streets and cheered. People threw flowers at his feet. A man named Vicente Azuero stood in the middle of the street. He began a flowery speech, extolling Bolivar as the greatest man alive. The Liberator interrupted, holding up his hand and saying, "No, no, I'm not the great hero you have painted. I'm a simple soldier. Please say no more." Pushing the man aside, he rode on, leaving the man startled and chagrined. A pretty young maiden, dressed in white presented Bolivar with a scroll and a wreath. He smiled broadly and kissed her cheek.

Another man stepped forward. "Great Bolivar, with one swift stroke you have freed Colombia!" The crowd cheered. Bolivar continued his triumphant tour until he dismounted at the Viceregal Palace. General Santander jumped down beside Bolivar, and they both received cheers of jubilation from the crowd. "The viceroy has fled," whispered Santander. "We are in complete control."

Bolivar turned to his friend. "Now, my dear Santander, I know for certain what I have always suspected. There is nothing I cannot do."

In the palace at Bogota the rooms and halls were large and austere. Furniture was scarce, paintings were faded and dust had settled in every corner. Yet, despite the air of neglect, the room bustled with activity. Aides-de-camp and secretaries regularly dispatched couriers, and petitioners came and went in a never ending stream. Through the main hall in a smaller room, Vice President Santander sat at a small desk studying rows of figures on long sheets of paper. Nearby, the president of Colombia, General Simon Bolivar, paced the floor, dictating letters and proclamations to three secretaries, each struggling to keep up. Santander looked up from his desk and shook his head. "Look, General, maybe we're going too far. We've taken all the money out of the Treasury. Remember, I'm Colombian. I feel responsible."

Bolivar waved impatiently. "I know, Francisco. We've taken 600,000 pesos in gold and silver. Besides that, we've reduced the salaries of government employees by half. We've confiscated the property of the Royalists. We've even appropriated the church's funds." He made a gesture of annoyance. "And, do you know what? It still isn't enough!"

"Don't forget the voluntary contributions from the rich," grumbled Santander.

"We need money for the war. Liberty doesn't come cheap."

Santander wiped his brow. "General, it was royal taxes that made the people turn against the Spanish. Dammit! The Spanish executed five hundred patriots, and nobody batted an eye. But when they raised the taxes, the people revolted and joined us!"

Bolivar shrugged and turned to continue his dictation.

"You've always had plenty of money," Santander shouted. "That's why you spend it like water."

Bolivar swung around and spoke firmly. "As you know, I used my entire fortune to finance the war. I spent it to equip armies. I gave it to the widows of my soldiers. Remember, the great George Washington also used his own money to pay his troops from time to time—because the Congress couldn't. Now, if you'll excuse me," Bolivar continued, "I'm finished with this conversation. It's beneath us, my dear friend."

As had happened many times before, Bolivar felt utterly alone. How could he make them understand that freedom was priceless? That it was worth *any* cost!

Although he wouldn't admit it, the daily routine of administering a government was beginning to weary General Bolivar. He hated placating the politicians, granting favors, having his ideals opposed at every hand. Everybody wanted his personal attention. No matter was too trivial for the Liberator's expressed consent and agreement. But he was a man of ideas and ideals. Minor details annoyed him.

To escape the bustle of palace activity, Bolivar continued his practice of taking two baths every day. Usually he bathed under waterfalls or in streams. Always, he enjoyed it. The other men, he knew, bathed far less frequently and viewed his daily cleansings as something of an obsession. But he didn't care. Bathing invigorated him.

He also arranged to have a small study just off his main office in the palace. The floor was littered with rough drafts of letters, yellowing government documents and a few old newspapers. Bolivar was sitting at a desk writing with his quill when Vice President Santander raced through the door, bristling with anger. "Sir! Sir! This decree of yours freeing the slaves—it's preposterous. The people will revolt."

Bolivar looked up with exasperation. "I don't care what they do. As I've said a thousand times: *Slavery is the daughter of darkness.* It's morally wrong." Then, he smiled. To placate Santander, he said, "Besides, the slaves make good soldiers. I need them for my army."

Santander remained visibly upset. "Sir, everybody knows you freed your own slaves years ago. Isn't that sufficient?"

"Absolutely not."

Santander took a menacing step forward. "It's that damned fool promise you made to President Petion of Haiti, isn't it?" Santander knew the story as well as anybody. Bolivar had arrived in Haiti with nothing, and he and President Petion, who had once been a slave, became as close as brothers, even though Bolivar was an aristocrat and former slave-owner. It was Petion who gave Bolivar the arms and ammunition he needed to continue the fight to free South America. Without this aid, Bolivar's cause would have been lost forever. Rumor had it that Bolivar had pledged to free the slaves in any territory he liberated.

Bolivar rose from his desk. "Listen, Francisco, this is my country now, and I can't waste my time arguing with you over every one of my policies!"

Santander clenched his fists. "How dare you, sir! I am the Colombian, not you!"

Pounding his desk, Bolivar shouted, "But *I* am president! It is my duty to free the slaves of Colombia. I will fight for this to the end! Then I will defeat Morillo! Remember what I said in my last speech: *I am the son of war!*"

In disgust, Santander pivoted on his heel and stormed across the room. At the door, he turned back and said bitterly, "Why don't you go all the way, General, and proclaim yourself the son of God!"

Bolivar winced as the door slammed, then sighed. Colonial rule has taught people that slaves are property, he thought, like land and money and horses. But they're human beings who deserve to be as free as I am. And they will be!

* * *

The next morning, Santander entered the Liberator's small office quietly, almost diffidently. He coughed softly to get Bolivar's attention.

The Liberator had been concentrating on several drafts of new decrees. Looking up, he nodded to his vice president. "Now, what is it?" he snapped.

"About our conversation yesterday," began Santander. "I've been thinking. I've been talking to people."

"The slaves will be freed. I decided that already. There will be no more discussion." Bolivar returned his gaze to the papers on his desk.

"Sir," said Santander, "our economies are based on the labor of so-called slaves. We have liberated only one country, Colombia. Venezuela is still ruled by Spain, in spite of your declaring Angostura its capital and the Congress electing you president."

"So?"

"It will wreck the productive agricultural capacity of Colombia, and the people will have no recourse but to revolt. The Venezuelans will certainly hear about your policy, and they will revolt, as well. There will be no more revolution."

Bolivar was now giving the problem his full attention. "Slavery is an abominable institution. I shall fight it with every means at my disposal."

Santander was pale, holding back his fear of being refused. He took a deep breath and said, "I have a plan, if you'll hear me out."

"Go ahead." Then, sensing his lieutenant's discomfort, Bolivar smiled at him and said, "Look, we're friends, Francisco. You know I'll listen to you. You're my closest colleague. I depend on you to give me good advice."

Still apprehensive, Santander said, "Well, let us say I accept your desire to free all the slaves in the countries we liberate."

Bolivar nodded curtly. "Don't forget Mama Hipolita was a slave. Yet she is the only mother I've ever known. And Jose Palacios."

"I propose this," said Santander slowly. "We shall, of course, free the slaves . . ."

"Good."

"But in a way that will not ruin our countries. First—and you will like this, sir—we will immediately free all male slaves over sixteen years old who will join our army."

"Excellent!" The Liberator showed enthusiasm for the first time that morning.

"And, second," continued Santander, "all children born of slaves from this day forward will be free."

Bolivar was pensive. "A gradual emancipation," he said softly. Then, he nodded. "Yes. I have been worrying a bit about freeing all the slaves at once. If I may be frank, I was almost dreading the chaos it would cause." He got to his feet and extended his hand to Santander. "It's a good plan. After one generation, there will be no more slaves in our country, only free men. The result will be the same, but the process will be much less wrenching. Yes. It is a good solution, Francisco."

Both men smiled. Both were pleased. A potential disaster had been averted, and they both knew it.

Chapter 25

With Bolivar's attention focused on Colombia, the Congress of Angostura was crumbling, threatening Venezuela's independence. To ease the situation, Bolivar made preparations to ride to Angostura with his new aide-de-camp, Daniel O'Leary.

Two days before they departed, Bolivar assembled his officers in the large presidential suite in the palace of Bogota to announce that Santander would be in complete control of the government during his absence. Despite his uneasiness, Bolivar declared his absolute confidence in the vice president. He had no choice;

Venezuela needed him, and he had to leave somebody with authority in Bogota.

The Congress at Angostura was meeting in the main hall of the building which was their gathering place. It was a large, bare, whitewashed room with a high ceiling and two crystal chandeliers. Through the door at the rear, Bolivar and O'Leary entered quietly and took in the scene unfolding before them.

The presiding delegate banged his gavel for order but the fevered crowd ignored him. One man stood on his chair and roared, "We'll banish Bolivar to Colombia since he likes it there so much."

"To hell with the tyrant Bolivar!" shouted another. "Let's write the constitution the way we want it."

The other members cheered in agreement.

"He wants us to copy the damned United States's system. To hell with it. Throw him out!"

"We'll give freedom to everybody! We won't have a president." Voices rose around the hall.

"We'll hang Bolivar," screamed a congressman.

O'Leary glanced over at his commander. Bolivar was rigid, seemingly stunned by such overwhelming, passionate opposition.

Motioning O'Leary to stay where he was, Bolivar walked confidently towards the front of the hall.

As he strode down the aisle, the men fell silent. "My god, it's him," whispered one. "He's here." Though Bolivar was a small man, the power of his presence was astounding. He mounted the podium, faced the assembly and waited until each member had taken his seat. They looked apprehensive. Several wiped the perspiration from their brows. The room had fallen absolutely silent.

Bolivar leaned forward and said in his high-pitched voice, "Where have I been, you might ask? Well, dear friends and colleagues, you must congratulate yourselves. By granting your president the authority to take the army into Colombia, you have enabled me to completely liberate your sister republic. I return to you triumphant. Every battle was a victory! Every soldier a hero!"

Bolivar raised his arms in the air. A few men applauded tentatively. Others were skeptical.

"What about Venezuela?" someone shouted. "You abandoned us!"

"Never!" the Liberator yelled back. "I secured your freedom. Liberty means nothing if your oppressors live next door! If the continent is not free, you are not free!"

One by one, the men began clapping, encouraged by their neighbor's mounting enthusiasm. Soon they abandoned their apprehensions and openly applauded their leader. Some even stamped their feet and cheered. Several raised their hands to be recognized. Bolivar pointed to one and made a motion with his hand to silence the others.

"Great Bolivar," the man began, "we were angry but now we welcome you back to your homeland. You have crowned our land with glory. You have freed our brothers and vanquished every foe. You have given us laws and stability. You are the greatest hero since the world began." His words caused a pandemonium of acclaim. One of those who cheered the loudest was the man who had just called Bolivar a tyrant and suggested he be hanged.

The delegates continued their eulogies, each praising Bolivar in more extravagant terms. Finally a man in the back called, "Make way! Make way!" as an army officer entered the hall. He walked slowly down the aisle until he reached the Liberator. "General Bolivar, sir, your generals are waiting on you outside." The hall was silent. Bolivar looked to the back of the room, his visage stern and intimidating. "They wish to submit to you, sir. They regret disobeying you, and they throw themselves on your mercy."

Bolivar nodded to the officer and descended from the podium. He walked toward the door, stopping to shake a few outstretched hands. Outside, he surveyed the small group of generals. They huddled together, each knowing that Bolivar could have them shot for treason. Instead, he gave each a warm embrace and chatted cordially. Many were on the verge of tears, overcome by Bolivar's clemency. One actually kissed his hand. A man appeared in the door of the congress hall. He took in the scene, then shouted, "Long live the great Bolivar! Long live Bolivar!"

The cry was echoed by those in the hall and rose to a crescendo that could be heard for miles around. *"Long live Bolivar!"*

Chapter 26

In Bogota, General Francisco de Paula Santander was the genial host at a luncheon in the large hall of the palace. At his long table was the former Spanish commander, young Colonel Barreiro, who had led the Spanish troops gallantly but fruitlessly at Vargas and Boyacá, with several of his other captured Spanish officers. The rest of the thirty-eight Royalist officers sat at other tables with Santander's officers. The wine was passed. Santander was laughing as he finished a rather risqué story. The Spaniards joined in the laughter. "That reminds me of one," said Colonel Barreiro. "Tell me if you've already heard it." He then launched into a hilarious joke in which the Spanish King was the butt of the punch line. Those at the table roared.

Santander slapped Barreiro on the back and said, "That was damned good!" The mood of the luncheon was cordial, almost festive, as if all the atrocities committed by the royalists in Colombia were forgotten. The wine flowed freely and a small orchestra began to play. At one point, Santander mentioned casually that the Liberator had offered the ex-Viceroy, Samano, an exchange of prisoners as a humane gesture, but Samano had not even had the courtesy to reply.

"Samano's no damned good," said Barreiro. "Let me tell you how he left town. Dressed as a woman, that's how. What do you think of that?"

"Just like him, the son of a bitch," said Santander.

"Here's a toast," said a Spanish major rising unsteadily to his feet. "Here's to Samano burning in hell!"

Everybody cheered and drained their glasses.

As enlisted patriot soldiers scurried back and forth refilling the glasses with wine, Barreiro turned to Santander and said, "You know, your Liberator has promised us a safe passage home to Spain. But before we go, how about letting us try to find that bastard, Samano, and turn him over to you? I'm sure you'll know what to do with him."

Everybody at the table laughed uproariously.

"Say, General. You could have him shot wearing his dress."

Santander laughed heartily. "If he showed up before the firing squad in a dress, I'm afraid they'd laugh themselves to death before they could shoot the bastard."

"Whose dress was it, I wonder," said a Colombian captain. "Maybe we should all go looking for a naked woman!"

"That's right," said another Colombian. "Our Spanish friends looted us so thoroughly nobody has more than one set of clothes to their name."

"Great!" said a young lieutenant. "The first one who finds her can keep her."

"I hope you find her and that she's old and fat," said his friend.

"That's all right," said the lieutenant, "as long as she's rich."

The officers all laughed. In one corner four Spaniards had formed a quartet and begun to sing.

When lunch was over, Santander rose and waved his hand casually. "Now, we have an appointment in the Central Plaza." He smiled.

The Spaniards gasped. Soldiers appeared and quickly fastened iron shackles on the lunch guests and led them away. In the plaza, the enemy officers were made to kneel, still shackled, and Colombian officers shot them one by one in the back of their heads. Watching the spectacle with great pleasure, Santander stood in the doorway of the government palace, from which the Spanish viceroy had ordered the executions of hundreds of Colombian patriots. A man rushed up to him. "General! The Liberator promised to treat the prisoners with leniency. He promised to send them back to Spain. He won't let you get away with this."

Santander turned to one of his adjutants. "Shoot him, too. He's obviously one of them." Without more ado, the man was dragged to the center of the square, forced to his knees and shot.

As the smoke began to blow away, Santander addressed the hastily assembled crowd. "I find an inner pleasure in having all Spaniards killed. These were the officers who butchered our fathers and brothers, our best friends and companions in this very plaza. Today, we have partly avenged those who died so miserably at the hands of these bloodthirsty Goths."

Then, preceded by the small orchestra, he rode around Bogota singing a song he had composed to commemorate the shootings. The song began by describing the officers in their chains with expressions of stark fear on their faces. It ended with the lurid scenes of the men in their death throes. Santander sang it joyfully, savoring the sweet taste of revenge.

In the kitchen of his small house in Angostura, Simon Bolivar was livid. "How could he do this to me?" he screamed. "He's made us look like savages in the eyes of the world!" He hurled a cooking pot against the wall.

"I think the fact he disobeyed your orders is much more serious." Danny O'Leary was worried. "Even though he claims the Spanish officers were about to lead an uprising."

"Nonsense. How could they? No, Santander is a terribly vengeful person. He despises the Spaniards for executing most of his friends in Colombia. He shot those officers in cold blood, and he knows it. In his letter, he even asks me 'To cover for him.'"

"Are you going to?"

Bolivar didn't answer immediately, even though he had regained his composure. "What Santander did is an accomplished fact. The Spaniards are already dead and buried. There is nothing I can do to change that. On the other hand, I want to consummate the union between Colombia and Venezuela more than anything in the world and without Santander's agreement to help I'm not sure I can."

"You're going to put political expediency before your personal feelings?" O'Leary clearly did not approve but Bolivar nodded slowly.

"I simply don't have time, Danny. I depend on Santander to keep Colombia under control and to continue administering my government. Besides, I think he's sufficiently frightened of me that he won't commit any more atrocities like this again."

Several weeks later Bolivar was sitting in his small garden in Angostura when O'Leary brought him the news: both the Congress at Angostura and the assembly in Bogota had approved his proposal for the union of Venezuela and Colombia, the united nation to be known as Great Colombia. They had also elected Bolivar president. But Bolivar was hardly jubilant. The news that his most ardent political ambition had been achieved brought him no joy for two reasons: The first was that, due to the inordinately slow communications, he had just learned of Pepa's death, which saddened him deeply, and the second was that he knew until he defeated Marshal Morillo and drove the Spanish out of Great Colombia for good, he would feel nothing but defeat. Nodding at O'Leary, Bolivar said, "We still have to beat Morillo. Until we do, all this means nothing."

Danny O'Leary looked worried. "Morillo's army outnumbers us in men, supplies, horses and guns."

"That's not all. He has more reinforcements coming from Spain."

"Then, we're beaten. There's nothing that can save us."

The Liberator smiled to calm his aide's fears. "Oh, yes there is, Danny. A miracle can save us. And I believe in miracles."

In the meantime, he would miss his Pepa. He felt a vacuum which he concluded could never be refilled in spite of all the women who threw themselves at his feet. Pepa had been exceptional.

Chapter 27

Marshal Morillo took the case from his aide and walked quickly down the marbled hall to his office. The aide followed him into the huge, sumptuously furnished room. Morillo opened the case and spread out the contents on his large rosewood work table. As usual, there were routine directives and orders, instructions and proclamations, which Morillo placed in neat piles to attend to later. The more important envelopes, he opened and read. After perusing one, he suddenly fell back in the chair. "Leave me," he said to his aide. Then, as the young officer was closing the door behind him, Morillo shouted, "Wait, come back!"

"Sir?"

"There's news. You must realize it's bad news, and I don't want you spreading rumors, so sit down and let me digest this thing."

The young officer sat down. Morillo looked out the window, deep in thought. When he turned to his aide he spoke slowly. "We won't be getting any reinforcements. The troops awaiting transportation to the colonies mutinied in the ports of embarkation. A couple of colonels led the mutiny and forced King Ferdinand to restore the liberal constitution of 1812."

"When did this happen, sir?"

"Three months ago. On January 1st." The year was 1820.

"No reinforcements?" The aide seemed in a trance. "None?"

Morillo rose to his feet. "None!" He leaned forward, cradling his stomach, and groaned. "Damn! That wound I suffered at El Semen still bothers the hell out of me. Damn European politics."

The aide looked puzzled. "I don't understand how all this happened. You reestablished Spanish rule five years ago, sir. That man, Bolivar, had to flee to Jamaica."

Morillo nodded. "Five years ago, the only rebel Spain hadn't beaten was San Martin in Argentina but we don't give a damn about Argentina." He frowned and continued. "Every time Bolivar was beaten in the north, he returned to fight again. In the south, San Martin was never defeated."

The aide hesitated. "But why doesn't Bolivar give up when he's beaten?"

"Ego! He's like a spoiled child. Can't stand to lose." Morillo picked up the dispatch case again. "I am to publish the 1812 Constitution and restore peace 'by fraternal conciliation.' Do you hear that? Fraternal conciliation. Impossible." He pounded his fist on the table.

"What does that mean, sir?"

"It means Madrid wants us to kiss and make up, just like that." Morillo shook his head. "I'm not sure we can. There have been too many atrocities on both sides." He looked at the orders lying on the table. "I shall do my best to obey, of course. But I think this means the end of Spanish rule in these colonies."

Simon Bolivar was smiling as he entered his office. Across the room, Captain Daniel O'Leary jumped to attention. He was still young and rosy-cheeked, brimming with boyish impatience.

"Sir, you seem unusually cheerful this morning."

Bolivar turned as though only just realizing O'Leary was there. "Of course I'm cheerful. Haven't you heard about my miracle?"

"No, sir."

Bolivar laughed. "There will be no more troops coming to America from Spain. Do you hear? My miracle happened."

O'Leary grinned broadly. "Does that mean we can beat Morillo, now?"

"Forget about beating Morillo. Morillo knows he can't win without reinforcements. There won't be any reinforcements so Morillo will want to make peace. Now is the time for *diplomacy,* my dear Danny. Let's go to the conference table with our wits about us."

Chapter 28

Marshal Pablo Morillo rode into the small town of Santa Ana on November 27, 1820. A muddy road ran through the village with a few one-story buildings on each side. Morillo wore his full dress uniform, decorations and sash. Fifty officers and a full squadron of cavalry accompanied him. The marshal seemed wary; his eyes darting in every direction. He turned to an officer beside him. "You sent scouts out to reconnoiter?"

The man nodded then pointed to the single rider approaching them from the west. He wore the field uniform of the patriot army and the insignia of a captain. Pulling up his horse before the Spanish contingent, he touched his cap and announced, "I am Captain Daniel O'Leary, aide-de-camp to President Bolivar. I've come ahead to inform you that the president is on his way and will arrive shortly."

Marshal Morillo rode forward toward O'Leary. "How large an escort does General Bolivar have?"

"Twelve officers, sir, including me."

Morillo nodded. "My old enemy seems to have outdone me in honor." He turned to his second in command. "Send back the hussars. We won't need cavalry today."

In a few minutes the small party of patriot officers appeared, all seemingly happy and at ease. After scrutinizing the riders, Morillo touched O'Leary on the shoulder. "Which one is Bolivar?"

O'Leary pointed to the smallest man in the party. "The man in the blue cloak."

Morillo shook his head in disbelief. "He's riding a donkey."

"A mule, sir. It's a symbol of peace."

Morillo dismounted and walked toward Bolivar on foot. Bolivar did the same and when the two men met, they embraced.

The next day, after a festive evening of toasts, good food and mutual admiration and camaraderie between Morillo and Bolivar they prepared to depart. The two men embraced again. Bolivar said, "Last night I toasted you as my noble adversary, my worthy opponent, my gallant foe. Now, for six months, I shall call you my virtuous friend."

Morillo, who had succumbed completely to the Liberator's charisma, responded, "You are most generous. And you have nothing but my greatest respect." The Spanish marshal paused. "I shall never fight you again, General Bolivar. I am going home."

They embraced once more. Then as each man mounted his horse and rode off in opposite directions, the cheers of both the patriot and the Spanish staffs rang in their ears.

Bolivar seemed unusually happy as he rode away from Santa Ana. Riding beside him, O'Leary ventured, "Things went well, sir?"

Bolivar patted his horse's neck with deep satisfaction. "Danny, my boy, my victory at Santa Ana surpasses Boyacá by miles!"

"But you and Marshal Morillo signed an armistice for only six months . . . . Of course! That gives you the time you need to rebuild your army."

"Yes," said Bolivar. "But that's nothing compared to this. Read the beginning of the agreement, my friend. It mentions by name the governments of Spain and *Colombia!* Spain has admitted to the world that we exist! *We are legitimate. Great Colombia has been born! And I am her father.*"

Chapter 29

Several months later, Bolivar set up his headquarters in a comfortable manor house on a large ranch in Venezuela. His troops were quartered in tents nearby. Even though the cavalry horses were pastured away from the house, the smell of manure permeated the air, and huge horse flies swarmed everywhere—indoors and out. A wide veranda surrounded the spacious building on all four sides. It was evening. Bolivar was, as usual, reading letters and dictating replies to a couple of secretaries on the front veranda. A hammock hung nearby; candles flickered in the breeze. Captain O'Leary galloped up to the front of the house and bolted up the steps, passing messengers and aides who were arriving and leaving. He held an envelope and his expression was grave. Bolivar looked up from reading what seemed to be a long letter.

"What news, Danny?"

O'Leary waved the envelope and declared, "The Spanish have refused to respect the armistice. They say we broke the terms when we entered Maracaibo."

"Maracaibo gave themselves up voluntarily to Urdaneta. He's from there. Do you have any answers to my letters to the King of Spain or to Marshal Morillo?"

"Nothing yet, sir. As you know, Morillo refuses to fight you and has returned to Spain. De la Torre has taken his place." O'Leary handed Bolivar the envelope he'd been holding and said, "Marshal La Torre refuses arbitration, sir."

Bolivar took the envelope and read its contents. He seemed sad and shook his head slowly. Finally he turned to his secretary. "I'll finish that letter I was writing to Marshal La Torre." He picked his words deliberately. "Say: 'It is my duty to my country to bring about a lasting peace—or fight.'"

There was no middle ground and it became obvious that the armistice would soon be terminated.

Simon Bolivar sat astride his charger with O'Leary at his side. From their vantage point on a slight rise they surveyed the Spanish entrenchments on the plains below. Their own troops were still winding their way into attack positions. In a dramatic gesture, Bolivar swept his arm across the horizon. "Before us lie the plains of Caraboba. What happens here today will decide the fate of an empire."

At eleven o'clock on the morning of June 24, 1821, the Battle of Carabobo began with General Antonio Paez's cavalry attack on the Spanish right flank. They were decimated by Spanish artillery, and from then on the slaughter on both sides was ghastly. Cannon balls flew, wounded men and horses thrashed, men shouted, horses reared, wounded men were trampled. At the patriots's darkest moment, drums rolled, bugles blared and flags snapped in the air. With bayonets flashing and colors flying, the British Legion advanced in perfectly straight lines. Even as huge gaps appeared in their ranks, the gallant Englishmen marched on. Colonel Farrier, their popular leader, went down mortally wounded, as did his second, Major Davis, and his replacement, Captain Scott. The Legion lost seventeen officers in fifteen minutes. But the bayonets of the Legion broke the Spanish line and turned the tide of battle.

The fighting became a series of infantry and cavalry charges. The gallant black colonel, Juan Jose Rondon, who had "saved the Fatherland" at Vargas led his thundering regiment against the royalist flank. Seeing Rondon riding ahead of his troops, leading the charge, his second-in-command, Lieutenant Colonel Mellado exclaimed, "In front of me, only the head of my horse!" and spurred forward. Passing Rondon, he threw himself headlong onto the Spanish bayonets, and was killed instantly.

Antonio Paez, in a decisive charge, led a hundred riders against a thousand Spaniards with such skill and daring that he routed them completely.

Covered with blood and the dust of battle, Uncle Antonio Paez rode slowly, observing the Spanish retreat. He saw Pedro Camejo disengage from the melee and start to ride towards him. Paez waited with a questioning look on his face. When his friend drew even with him, Uncle Antonio welcomed him, then joked, "I always thought you were the bravest of the brave. Why are you leaving the battle? Are you suddenly afraid? Lost your nerve?"

Pedro Camejo shook his head slowly. "No. I've come to say farewell. I'm killed." With that, he slid off his horse.

Paez jumped down and turned his comrade onto his back, but as he held his head in his hands, he saw the lifeblood of his faithful friend flowing onto the ground.

The enemy showed equal valor. Marshal de la Torre led one regiment after another in suicidal attacks against the patriots.

After Bolivar's staff had thrown themselves into battle as courageously as the line troops, O'Leary galloped up to a group of leaderless infantrymen and leapt from his horse. He grabbed the arm of the nearest soldier. "Where are your officers?"

There was consternation in the man's eyes. "Killed, sir. All killed."

"Steady, men!" O'Leary's voice could be heard above the roar of battle. "Form ranks. Back into your squads."

The soldiers moved quietly into organized bands. Although clearly frightened, they obeyed. "Form on the right flank of the Rifles Battalion," shouted O'Leary, running in that direction. The men followed him and formed ranks. "Prepare to receive the enemy," he shouted. The men loaded, cocked their rifles and fixed bayonets.

O'Leary looked to his front. The enemy infantry was coming toward them, picking up their pace steadily as they closed on the patriot line, but they hadn't fired yet. Suddenly, a prematurely discharged musket ball smashed into a patriot soldier's arm. The man didn't even flinch, but when he raised his arm, O'Leary could see a bone piercing the man's uniform and blood spurting from his arm. Suddenly, the man's face contorted with fear and anguish. Holding his arm, he tumbled to his knees and rolled over, writhing in pain.

O'Leary was already breathing hard when the full force of the Spanish volley struck his thin line. Many of the men went down, hurled to the ground by the large musket balls, dead in an instant. A man was hit in the throat causing his blood to spurt out like water from a ruptured fountain. Another put his hands to his face, only to find nothing but a mass of blood, teeth and remnants of an eye. Another lay squirming in the dust, clutching his crotch; blood trickled through his fingers, staining his trousers a bright red.

While the wounded presented a shocking tableau of flesh in torment, their companions in the ranks stood tall, faces expressionless, eyes directed at the enemy.

"Ready!" shouted O'Leary, as the enemy charge closed in.

"Aim!"

The Spanish were now running towards the Colombian line with fixed bayonets, their shot expended.

"Fire!"

In a flash, the patriot rifles roared. Their balls stopped the Spanish charge like a brick wall. Now it was the enemy who writhed and screamed in misery. O'Leary genuflected and whispered, "Lord, if I have to go, please kill me quick."

He raised his saber high over his head and shouted, "Forward! Charge!" His men rushed into the reeling Spanish line to bayonet the enemy soldiers before they could recover from the shock of the patriot fusillade. Stabbing, jabbing, using rifle butts to smash heads, the men fought quickly and effectively. The ground became slippery with blood and gore. The few Spanish survivors began to run away from certain death as patriots moved forward.

For the moment, the enemy thrust had been repulsed. The Colombian line had held.

The tide of battle swayed back and forth, breaking up into a series of bloody skirmishes and hand-to-hand combats. The dead and wounded piled up on the plain. The losses were ghastly in both armies. Bolivar saw Marshal de la Torre withdrawing his remaining men. But there was still massive confusion—men running, horses thrashing. Any plan or coherence was disintegrating on both sides. The battlefield was absolute chaos.

Bolivar spurred his horse into the fight, shouting, "Order! Discipline! Form ranks! Form ranks! Officers, take charge! Regroup! Regroup!" Gradually, the lines reformed, and the patriots began an orderly pursuit of the surviving Royalists. Bolivar's victory was complete.

The next day, Bolivar sat in his tent and dictated a letter to Santander: "Yesterday, I came to the plains of Carabobo with the finest Venezuelan army ever to bear arms. And by their splendid feats in battle our soldiers assured the political freedom of the republic. Victorious, all Venezuela is now free."

Chapter 30

While Bolivar was liberating Colombia and Venezuela, the Great Argentine, General San Martin, had won victory after victory in Argentina, crossed the Andes with his cannon and his cavalry and helped liberate Chile as well. Now, he had entered Lima in Peru. The Spanish viceroy, La Serna, sensibly retired with his army and that of Marshal Canterac to the High Andes to await developments. He knew San Martin did not have the force necessary to follow him into the mountains, nor was San Martin's political power or acumen sufficient to hold onto Lima for long without help.

Encouraged by San Martin's taking Lima, the coastal city of Guayaquil declared its independence.

The Liberator had the capital of Great Colombia set up in Bogota because he always considered that Ecuador would constitute the third country of Great Colombia, and Bogota was central to both Caracas and Quito. The Spanish still held Quito and could descend on Guayaquil and quell the rebellion at any opportune moment. For this reason, Bolivar sent General Antonio Jose Sucre to Guayaquil with a thousand men. He arrived just in time. The Spanish had decided to attack Guayaquil and bring the city back into the colonial fold. With his small force, Sucre defeated the two large Spanish armies sent from Quito to take Guayquil. He attacked and routed the first, then turned and sent the other fleeing back to Quito. Against Sucre's advice, the Guayaquil government sent its army into the mountains to take Quito. They were soundly beaten and had to flee for their lives.

Bolivar had promised the people of Quito that he would come liberate them. They had risen before but had been subdued by Spanish arms and knew they hadn't the strength to succeed in another effort. To keep in close touch with San Martin's intentions, Bolivar named Joaquin Mosquera his Emissary Extraordinary to Peru, Chile and Buenos Aires, with instructions to remain in Peru and keep in touch with San Martin. The Liberator was determined to go south to achieve his goal of freeing Quito and the rest of Equador, including Guayaquil, even though he knew the Spanish had large armies both in Quito and in Pasto, farther north.

But before Bolivar could go south to liberate Ecuador, as he had promised, Panama had to be brought into the Republic of Great Colombia as well. Bolivar had prepared to send his boyhood friend, General Mariano Montilla, with a small army to drive the Spanish from that province when, on November 28, 1821, the people of Panama rose against Spain and declared their independence. The leading men of Panama then announced their desire to join the great Bolivar and become a part of Colombia. This was a magnificent coup. With Panama united to Colombia, Bolivar could now send armies south by ship from the isthmus. But even more important, the Spanish colonies on the Pacific were cut off from all reinforcements from the mother country.

Chapter 31

The Andes Mountains were a natural fortress for Ecuador. In the north, they created a stronghold at Pasto which had shattered every attack the patriots attempted. The walls of the Andes also protected Quito and its strong Spanish garrison composed of seasoned combat troops ready to move in any direction. The Spanish had no fear of losing Ecuador. The Mountains were their allies. They could take back the coastal city of Guayaquil whenever they pleased—once Sucre had departed.

To draw Spanish troops away from the defense of Quito, the Liberator moved towards the city of Pasto which was solidly Spanish. At midday on April 7, 1822, Simon Bolivar's troops filtered down a small slope of the Andes near Pasto and crossed a stream in orderly columns, one regiment following another. By the stream, Simon Bolivar sat astride his charger, surveying the situation. Spanish soldiers lined the hills in front of him. His friend, General Pedro Leon Torres rode up, saluted and pointed at the Spanish troops. "Sir, the enemy's positions are impregnable."

"Yes. But we can't remain where we are, and we can't withdraw. So, we shall attack. And we shall win."

"What's this place called?" Torres continued to survey the Spanish troops on the hills.

"Bomboná. Why?"

"Because it's going to be a damned bloody place pretty soon. And we're going to have to tell a lot of widows and mothers where their husbands and sons died. Bomboná, you say?"

A flash of irritation crossed Bolivar's face as he turned to Torres.

As his other officers arrived for their orders, the Liberator kept reevaluating the situation. Finally, he announced, "General Valdes will take his division, plus Major Sandes' Rifles Battalion, up the hill to surround the left flank of the Spanish." He looked directly at Valdes. "It will be the most demanding thing you have ever done. You will have to get up and around that rocky hill that protects their entrenchments and approach from the rear. It's our only chance of beating them."

Valdes nodded. His face was grim.

"General Torres will attack the Spanish center without delay. It's late. Let's get going." Bolivar continued to sit his mount as his generals returned to their respective units.

After he had moved his troops into position, General Torres turned to his second in command. "Let the men rest a bit and eat lunch. They'll need it. I think the Liberator wants Valdes to show some progress before we go into the attack."

"General Bolivar said, 'without delay,' sir."

"He meant after Valdes's division gets up the hill. As soon as they've surrounded the Spanish left, he wants us to attack without delay."

With an aide beside him, Bolivar watched Valdes's men climb the hard rock on the Spanish left. Bolivar seemed impatient and nervous. "Dammit! Where's the attack on the Spanish center?"

"I'll go see, sir," suggested his aide, a young colonel named Jose Gabriel Perez. Danny O'Leary had been detached to serve with Sucre.

"No. We shall both go."

When the two men reached Torres's division, they found most of the men sitting on the ground finishing their lunch. General Torres came over to greet his commander. "Good afternoon, General. How is Valdes doing?"

"I ordered you to attack!" Bolivar screamed. "You have disobeyed me. Why? Why are you still here?"

Torres fell back a step, dismayed. "Sir, I misunderstood. I thought —"

"Attack! Those were my orders! Look at you! Look at your men! Lolling on the ground like schoolgirls on holiday, while we fight for our lives."

"But, sir . . ." began Torres.

"Don't speak. Brave men are dying because you refuse to act. I relieve you of command."

Torres stood white faced before his chief. His posture and contorted features betrayed his utter mortification. There was no use trying to explain. He had misunderstood and in doing so, had betrayed his leader and his men. As Bolivar glared at him, Torres drew his sword from its scabbard and broke it dramatically over his knee. "If I am not worthy to serve my country as a general, then, I will serve as a soldier. I shall join the ranks."

Simon Bolivar sat his horse in stunned silence. Remorse and shame showed clearly on his face. Without a word, he jumped down, threw his arms around Torres and embraced him. Tears filled Bolivar's eyes. When the two men pulled apart, Bolivar said, "I reinstate you to your full rank and titles. You are a brave man. I am sorry I spoke to you that way. You didn't deserve it. Forgive me." Again the two men embraced.

Later that afternoon on the battlefield of Bomboná, General Pedro Leon Torres gave his life fighting for Simon Bolivar.

The Battle of Bomboná was in furious progress. Cannon and rifle fire echoed again and again through the mountains. A rider reined up beside Bolivar, touched his cap and cried, "Sir! Sir! General Valdes has made it. He and Sandes are attacking the Spanish rear."

Bolivar acknowledged the salute. "Tell him, 'Well done.' But advise him we're not making any progress on the Spanish center."

Beside Bolivar, Colonel Perez sat in silence, overwhelmed by the slaughter he was witnessing. Quietly he said to the Liberator, "General Torres was killed leading the first charge."

"I know." There was an unusual sadness in Bolivar's voice. "And the first charge failed. I have thrown three more divisions at the enemy with no better luck. Their cannon and rifle fire are blasting our men to pieces."

Perez shook his head without replying.

"You see how bravely my men go into battle," said Bolivar proudly.

"Yes." Perez's voice was choked with emotion. "They go bravely—but they do not return."

As darkness fell, the battle continued. The soldiers on both sides were determined to fight until they were all dead. Only the blackness of night ended the carnage as the men could no longer see each other among the rocks and gullies. In the end, it was the Spanish who retired, leaving the battlefield of Bombóna to Bolivar. His losses had been frightful. Yet the men who survived were still a united army. In contrast, the enemy had scattered and were no longer a military force.

The next morning, Bolivar stood near the battlefield of Bombóna. Burial parties were extremely busy. The medical tents could not hold all the wounded, and many lay on blankets in the open field waiting to get their festering wounds bandaged. Squads of soldiers were out collecting the rifles and ammunition that littered the landscape, some still clutched in the hands of their dead comrades. Colonel Perez was reading the casualty reports to a dejected Liberator. "We lost over half of our men," Bolivar said with a husky voice.

"Yes, sir. But the Spanish lost the battlefield. So, all in all, sir, both sides lost this battle."

Bolivar didn't argue the point. He was worried. Their losses were devastating. The carnage, the killed, the maimed—it had all been appalling. Now he prayed that it had all been worth it, that they had pulled enough of the Spanish away from Quito to give Sucre a chance of winning. Either all was in vain—or they had won the liberation of Equador!

While Bolivar was drawing off the Spanish forces at Bombóna, his finest general, Antonio Jose Sucre moved his army out of Guayaquil to attack Quito. The patriots had tried at least five times to take Quito and in every instance were thrown back with heavy losses. Sucre had with him his Venezuelan regulars and the British

Legion plus a thousand troops sent by the Argentinean general, San Martin.

General Antonio Jose Sucre de Alcala gathered his generals and colonels, about ten in all, in a wooden hut behind his lines below Quito. In contrast to the uniformed men, he wore a pair of wool pants and a poncho. He was short and slender, like Bolivar, although much better looking, with luminous brown eyes and black hair. His face was thoughtful, more like that of a gentle poet than a man of war. Yet it was in war that his true genius lay.

Sucre was a young Venezuelan from a large, wealthy family whose ancient titles of nobility descended from European aristocracy. His forebears had been military leaders for generations, and as a cadet he studied military tactics and strategy with the best Spanish professionals in Venezuela. In 1812, at the idealistic age of sixteen, he joined the patriot army. His fourteen-year-old sister had thrown herself to her death from the family's balcony to avoid capture by Royalist renegades, an act typical of the bitter wars of independence.

Sucre fought bravely in the front ranks until the defeat of Miranda's republican army in 1812. Then, during the Spanish reprisals when many of his family, including his stepmother, were killed, he fled to Trinidad. He returned to fight at the head of his troops in Colombia until the conquering Royalists forced him to flee again.

He looked at the expectant faces of his officers standing in the hut and suddenly felt inadequate. He knew every one of these men respected him as an outstanding general; still he was the youngest man there, with the exception of Colonel Cordoba of the Magdalena Regiment. Many of his top officers were British and did not speak Spanish well, except for Danny O'Leary whom Bolivar had sent as his personal representative to Sucre's army. And although he had had a thorough military training and was a veteran of many battles, this was the first time Sucre would command an entire army on his own.

He took a deep breath and straightened his shoulders, his fleeting apprehension invisible to the men.

When he spoke, his voice was firm. "The Spanish hold the high ground. They expect us to attack them from our present positions below them and to massacre us as they have always done in the past. It has become a habit with them, a routine operation. However, gentlemen, I have reconnoitered the area . . ."

He paused and directed his aide to unroll a large map and nail it to a plank on the wall of the hut. "You will note that the heights of Pichincha lie at the rear of the Spanish army. I propose to make a forced march tonight around the enemy's flanks and attack them tomorrow morning from the heights at their rear."

"The men won't like it," said the British commander. "They're tired and it's going to be a hell of a march. All uphill."

Sucre smiled and sat on a nearby table, his leg dangling over the side. "Tell me, Colonel. Do you think they'd rather march tonight or get their heads blown off tomorrow morning by the Spanish cannon?"

The Colonel nodded. "What time do we start?"

When dawn broke, the Spanish army was stunned to find the patriots drawn up in battle order at their rear. The battle of Pichincha began at 10:30 on the morning of May 24 and was furious, hard-fought and decisive. Although badly outnumbered, the Colombians, Argentines and the troops of the British Legion battled courageously, sweeping the desperate Spaniards before them. Victory was finally achieved with a storming attack by the young Colombian Colonel Cordoba, leading his Magdalena Regiment. His triumph complete, Sucre accepted the surrender of the Spanish and declared all Ecuador free.

Chapter 32

Judging by its appearance, the city of Quito had been preparing for days for the arrival of the great Bolivar. Arches of flowers had been erected on every block and flags flew from every house and public building. Masses of people were out in the streets, all dressed in their finest, with ribbons of red, blue and gold, the colors of Great Colombia, fastened to their dresses and suits. For some, their most elegant clothing was little more than knee britches and metal shoe buckles worn by their fathers in the 1700's. Some women wore faded silk or velvet gowns handed down from the women they worked for. The mood was joyous. A group of young girls dressed as angels chattered and laughed as they waited to welcome the demigod Bolivar to their city. Everyone carried blossoms of flowers, brightening the city with their color and scent.

Interrupting the buzz of anticipation was the cry of a young man running urgently down the main street. "He's coming! He's coming! The great Bolivar is coming. Citizens of Quito, prepare!" An expectant hush fell over the crowds. Everybody strained to get a glimpse of General Bolivar, the hero of the world. An old man sighed aloud, "The great Liberator is actually coming to Quito." At the far end of the avenue, the cheers swelled into waves of joy as the Liberator rode in. Church bells rang jubilantly. The little girl-angels danced in front of Bolivar's horse, strewing his path with flowers from their gilded baskets. A military band struck up a march. Fireworks soared overhead. A state of delirium swept through the city.

From his mount, Bolivar basked in the adulation, smiling radiantly. He lifted his hat and bowed to the crowds as he passed; he waved and blew kisses at the pretty girls; he respectfully saluted the men.

General Sucre and his staff were waiting for him at the center square. Bolivar dismounted amid applause and cheers and embraced Sucre. Together they turned and waved to the people of Quito. Bolivar tried to begin his speech, but the crowds were irrepressible. Still smiling, the Liberator remounted his stallion and continued his triumphal procession, stopping every few minutes to rear his horse, eliciting a roar of approval from the crowd.

On the balconies of the larger houses, the more prosperous families waved excitedly and threw wreaths at the feet of the Liberator's horse. The women were beautifully dressed, the men well-groomed and formally attired. As they were obviously the gentry of Quito, Simon Bolivar paid them special attention, waving and tipping his hat. Abreast of one of the largest houses, he was preparing to pirouette his stallion for the benefit of those on the balcony, when something hit him full in the face. It was a laurel wreath that had been swept up in a gust of wind and carried off course. The Liberator lifted his head angrily to see who had tossed it.

On the balcony, a dark-haired young beauty stood among the spectators clutching her throat in horror. Her alabaster skin blushed rosy red; her brown eyes widened. Bolivar's scowl slowly turned to a smile. He lifted his hat and made a courtly bow, then reared his stallion. When the horse touched down, the Liberator searched the woman's eyes hoping she felt the same sudden surge of desire he did.

Then the cheers of the crowd rose again in Bolivar's ears. Glory beckoned.

That evening, Simon Bolivar and his staff dined at the mansion of Don Juan de Larrea, the patriarch of Quito society. They were met at the entrance by a long line of Indians in livery, each holding a torch to light the walkway to the front door. They wore powdered wigs and tricorn hats, starched neck stocks and heavy silken long coats, satin waistcoats and white velvet knee britches. Their flawless attire could have passed muster at any court in Europe—except that every single one of them was barefoot.

An hour later, in the vast second-floor ballroom of the Larrea mansion, the Victory Ball was in full swing. At the door, the party's host, Don Juan de Larrea introduced Bolivar to the arriving guests as "the Liberator of Colombia, Venezuela, Panama and Ecuador." Simon Bolivar greeted each new arrival as if he were the most important person in his life. And each moved down the receiving line with a rapt look on his face as if he had discovered that Simon Bolivar was his most cherished friend. Later, many who had just met him for the first time were heard to say, "Oh, yes, the Liberator and I are like brothers."

Actually, he longed to leave the receiving line and join the party. Glancing over his shoulder, he could see the women in bright-colored ball gowns swirling to the rhythm of the orchestra, gliding in and out of the arms of their partners—officers in full dress uniform with glistening gold braid.

Finally Bolivar turned to his host and asked, "When can we end these formalities? I like to dance too, you know."

Don Juan smiled benignly then turned to the next guest. "Señora Manuela Sáenz de Thorne, I'd like you to meet the Liberator of Colombia, Venezuela, Panama and Ecuador." Bolivar smiled perfunctorily then realized he was looking into the eyes of the woman who had hurled her wreath at him that morning. His face brightened; he bowed to kiss her hand. When he looked up, Señora Sáenz was returning his smile with a roguish grin and a twinkle in her eye. Their gaze was hypnotic and, for a fleeting moment, the room, the music and all the guests faded away.

Don Juan coughed discreetly. Manuela took a sudden breath, then nodded coyly and continued down the line.

After greeting several more guests, Don Juan turned to the Liberator and said, "I think we can break up the receiving line now," knowing that Bolivar would welcome the suggestion. "But for the sake of protocol and courtesy, I think you should dance with several of the prominent ladies of Quito." He took the Liberator's arm and steered him across the room. As they passed the punch bowl, Bolivar noticed Manuela joking with a group of his British officers. Colonel William Fergusson was laughing loudly, obviously having sampled more than a few glasses of port. Tall, slim and greying at the temples, Fergusson looked every bit the stately gentleman he was. Bolivar liked his company and trusted him.

Bolivar turned to his host, "You've met most of my officers, I think."

"Most, I believe. General Sucre is your white knight, of course. None better than he. I also think he's becoming interested in the little Marquesa de Solanda. Charming girl. I would feel proud, naturally, if the great General Sucre married a young lady from Quito."

Wishing to remain in the vicinity of the punch bowl and Manuela Sáenz, Bolivar asked, "And the others? Have you met them all?"

Larrea nodded. "Yes, but, tell me again. Who is that handsome blond Englishman over there entertaining the ladies so gallantly?"

"That's Colonel Arthur Sandes. He commands the Rifles Battalion which is the most disciplined and bravest unit in the army. Bolivar paused. "In the last four years, 22,000 recruits have gone into the Rifles Battalion. When it entered Quito, it numbered six hundred men."

"You mean that one small unit has lost so many men in just four years?"

Bolivar inclined his head. "The casualties in the infantry are always appalling."

Beyond the punch bowl, Larrea asked about two officers standing alone in the corner. They appeared to be in a serious discussion.

"The sandy-haired one is Colonel Rupert Hand," Bolivar answered. "An idealistic young Scot. In battle, he's a real professional—efficient, dependable, follows instructions to the letter. As a captain, he was wounded badly at Carabobo but recovered nicely. He's a bit of a Puritan, though. I suspect he doesn't quite approve of us Latins. He's far too straight-laced and proper."

"And the other?"

Bolivar smiled. "That's the hero of Pichincha, Jose Maria Cordoba." He waved towards the handsome, dark-haired young man, whose uniform sparkled with gold braid. "He doesn't know what fear is. He's twenty-three years old and has just been promoted to general. He loves war as passionately as most men love women. They make a strange pair. Cordoba is reckless and hot-blooded; Hand is disciplined and cool under fire; he's probably never disobeyed an order in his life." Bolivar stopped and looked around the room, eager for a diversion. "Now, my dear Juan, which of these lovely ladies am I to dance with?"

An hour later, he finished dancing with the niece of Don Juan, the last of his designated partners. He smiled and bowed. "Thank you, my dear. It has been entirely my pleasure."

She blushed and curtsied. But already Bolivar's eyes were scanning the room for Manuela. He finally spotted her on the other side of the room, standing alone by an open window. She was looking squarely at him, and he knew she had been waiting all night for him to look her way.

He moved toward her, feeling oddly nervous as he approached. But when he reached her, her smile was open and bright and dissipated his anxiety.

"Good evening again, madame. May I have the pleasure of this dance?"

"It would be all my pleasure, Your Excellency."

She glided into his arms, and together they danced effortlessly to the music. In the custom of the day, Bolivar was careful not to pull her too close, but was acutely aware of every sensuous touch their dancing allowed. Manuela was a skilled and graceful dancer, moving with both freedom and stateliness. Bolivar held her with pleasure and discreetly allowed her to lead him through the waltz. Other couples on the dance floor looked on openly, clearly more interested in the new twosome than in their own partners.

"You are very beautiful," Bolivar said. "And quite a marksman."

Manuela looked puzzled.

"The wreath," he explained. "I believe you hit your target. A bull's eye, in fact."

"Oh, I am sorry about that. I hope I didn't hurt you."

Bolivar twirled her adroitly then pulled her back into his arms, his expression grave. "I'm afraid the wound *is* deep. If I am ever to heal, I will surely need a great deal of care and attention."

Manuela smiled mischievously. "I see. And what sort of care and attention did you have in mind, your Excellency?"

"Hours of it."

She bowed her head obediently. "Since I am responsible for your wounds, I shall be pleased to see that you recover, no matter how long it takes."

"I think a stroll in the gardens would do me good. Perhaps there I can catch my breath of which your beauty has robbed me so completely."

Manuela inched ever-so-slightly closer and whispered, "Your Excellency, I don't intend to ever let you catch your breath again."

"Juan, they're gone!" Señora de Larrea was looking around the room and anxiously tugging on her husband's sleeve.

Don Juan de Larrea shrugged indifferently.

"Couldn't you see how fascinated he is with her? He's completely bewitched."

"My dear, Manuela Sáenz is a charming lady. And he's the idol of the continent. The enchantment is mutual. Let them alone."

The dance floor was still a vision of twirling gowns and immaculately pressed uniforms. Small clusters of partygoers were scattered throughout the ballroom, their voices inaudible beneath the small orchestra's music.

"I don't have to leave them alone, Juan. They are alone! And everyone's talking about it!"

The next morning, Simon Bolivar and Manuela Sáenz awoke together in the palace bedroom where the former Spanish governor had slept. It was a huge and majestic room with velvet curtains and a canopy to match. As the two lovers stirred, Manuela untangled herself from his embrace and propped herself up on her elbow. "So, it's true what I've heard; you're as ravenous and virile as a Greek God."

Bolivar gently pushed her back on the pillow and pressed his lips to her ear. "And you are more desirable than Aphrodite," he breathed.

Their lovemaking consumed most of the morning. Then, as they were dressing, Manuela asked, "Will I ever see you again?"

"Do you want to?"

"I want you to hold me in your arms every minute of every day."

Bolivar reached across for her hand and asked, "How about tonight?"

She lowered her eyes and said demurely, "Yes, my Liberator."

"Are you concerned that everybody will know?"

Manuela laughed. "They know already."

"That we've made love? Do you think so?"

"Not only that we've made love, but how many times. This is Quito, you know. It's my hometown. But what does it matter?"

"You're an interesting woman," Bolivar commented. "Most ladies are very concerned about these matters."

"Simon, I have been the source of rumors and scandals all my life. My mother was an eighteen-year-old spinster; my father had a wife and four children. They all found my very existence an enormous embarrassment and a considerable inconvenience."

Bolivar sighed sadly.

"When I was seventeen," Manuela continued, "I was sent to a convent to minimize the disgrace. My relatives shunned me. And I them. So, you see, I have long since stopped worrying about what the world thinks. I do not live by their rules."

Bolivar nodded thoughtfully. "I've often said the same thing myself."

"Yes, I know," Manuela said quietly. "We are the same."

He looked up quickly. This woman already knew him too well. She seemed to understand him better than those who had known him for years. He changed the subject. "So tell me, who's Thorne?"

"My husband."

He was pulling on his boots but stopped and looked up abruptly. "Your husband? Well, there's some news. Aren't you worried about him? If all Quito knows we're lovers, he's sure to hear about it."

"He's in Lima. Besides, I live my own life."

"Doesn't he care?" asked Bolivar.

Manuela shrugged. "We're not close; I don't honestly know him well."

"Why did you marry him if he means nothing to you?"

"My father was living in Panama at the time, and he sent for me. I didn't know it then, but he'd already made arrangements for me to marry James Thorne. And why not? It was a suitable arrangement for a girl 'in my position.' He's a nice enough man and very rich. Does it bother you that I'm married?"

"No. The first girl I ever copulated with was married. She was the wife of my tutor but not much older than I. I was fourteen." Bolivar shook his head. "But this rich man you're married to? What of love? How can you live without love?"

She moved across the room and took him in her arms. "Now that I've met you," she said softly, "I don't intend to."

Chapter 33

On the floor below, Major Daniel O'Leary was preparing the Liberator's temporary headquarters in a large corner room of the palace. The study had been cleared of its furnishings to accommodate Bolivar's staff and equipment. Now, in the bare, empty room, O'Leary was unpacking trunks of files, letters and petitions that Bolivar always carried with him. Nearby, several clerks were sharpening quills and arranging papers on empty boxes, preparing to transcribe the letters and proclamations that Bolivar was sure to dictate the minute he entered the room.

The door opened, and they all looked up, expecting the Liberator—but instead Rupert Hand walked in. He seemed unusually flustered this morning, waving a curt greeting to the clerks, then strolling purposefully over to O'Leary. "Damn it, Danny, you're supposed to keep this sort of thing from happening."

O'Leary smiled easily. "Good morning to you too, sir. Now, what seems to be the problem?"

"You know exactly the problem."

"Oh, come now, Hand. You know our Liberator as well as I do. What do you expect?"

"I expect him to behave himself in formal situations like last night. I expect him to keep his affairs private. For heaven's sake, Danny, he doesn't have to flaunt them the way he did. After he'd finished dancing with his assigned partners he took Manuela on the dance floor, gazed into her eyes and suddenly they were gone! Never came back for the rest of the evening. It was shocking! And the other ladies were hurt, I can tell you."

"You're too proper by far, Hand."

"Well, the whole town is talking. It's just like home. Everybody knows everybody else's business. This Manuela—her mother and her father were nobility, the highest in the land. Right? The only trouble was that her father was married to somebody else and her mother wasn't married at all!"

O'Leary couldn't help laughing. "Sure. It happens all the time. Why, Rupert, in the old country . . ."

"Forget the wretched old country."

"Well, it's not the girl's fault her mother and father weren't married."

"No, that's true. But her mother spent the rest of her life in a church praying for forgiveness. Do you think this Manuela appreciated all that? Hell no. The first chance she got, she ran off with a hussar."

O'Leary nodded. "Yes, I guess I know the story better than you do, Rupert. I'd just hoped word hadn't spread yet."

"Then you know the hussar dumped her? Somehow, her father got her married off to this man Thorne, who nobody knows anything about."

"He's British, like us."

"He's English. I'm a Scot. And you're a bloody Irishman."

"We're both Colombians now. And republicans, too. Except that you're still impossible, just as you always were." O'Leary smiled good-naturedly.

But Rupert Hand was not deterred. "Thorne's a rich English merchant. He lives in Lima. He's a good twenty years older than Manuela and owns a bunch of ships. He's crazy about his wife." Hand paused to catch his breath. "And his wife's in bed with our Liberator!"

Several days after the Liberator's arrival in Quito, young General Sucre quietly entered his office. The Liberator, who stood looking out a window, sensed his presence and came to greet his friend.

"Please sit down," he told Sucre though he remained standing himself. "The Congress in Bogota has empowered me to deal with these newly liberated lands as I see fit."

"You can thank General Santander for that. He knows how to handle the Congress."

"You're quite right. Anyway, I'm naming you the president of the Department of Quito."

Sucre rose from his chair. "Oh no, sir. Not me, sir. I'm a simple soldier; I don't know anything about being a president. I'm much too inexperienced; I don't . . ."

"Enough! You're far too modest, Antonio Jose. You'll make a fine president. You have ability and good common sense. You're a born statesman and leader. Your only problem is you're too kindhearted and gentle, but the masses worship you, so you won't have any problems there."

Sucre still seemed reluctant.

Bolivar was uncharacteristically impatient with his brilliant young friend. He'd sent Jose Palacios to bring Manuela to the palace, and he could picture her already lying naked in his bed. He was in a hurry to get to her side, afraid she'd get tired of waiting and leave.

"Besides," he snapped, "as the president, you'll be able to

court the lovely Mariana de Solanda. And, tell me, my friend, what girl can say no to a president?"

"I'd rather win her on my own merits. I want her to love me the way I am, not because I'm the president."

"Oh, for God's sake, man! You're going to be president and that's that!"

"Yes, sir."

Sucre smiled in spite of his protestations. His self-doubts had been resolved by the Liberator's decision.

After Sucre left, Bolivar wondered again how long the man would continue to be so plagued with self-doubt. Sucre had succeeded in virtually everything he had ever done. Perhaps, Bolivar thought, it was because he had lost most of his family and felt unsupported in the world. It didn't help that so many people said that Bolivar was jealous of him. Why does he believe these idiots, Bolivar asked himself. *He knows I'm his friend.*

Later that evening, Bolivar gazed out at the bright, equatorial stars shining like crystals in the night sky. He breathed deeply then returned to his bed where Manuela lay waiting.

Bolivar smiled. "Manuela, you fill me with passion! I just look at you and feel the stirrings."

"You know, my great Simon, you're the only man who's ever satisfied me."

He wrapped his arms around her and kissed her hard. Soon their bodies were moving rhythmically together, their breath coming in short gasps until, temporarily spent, they both lay quietly on the same pillow. Manuela stroked Bolivar's cheek. "How long will you stay in Quito?" she asked. The question hung in the air.

"Only another few days, I'm afraid. I have to go to Guayaquil. I've got to face General San Martin and have a showdown. An awful lot is at stake—in fact, nothing less than the future of South America. Damn! I wish somebody could tell me more about the man."

"General Jose de San Martin? I know him well, my dearest. Take me with you, and I'll tell you everything I know."

Bolivar sat up straight in the bed. "You know San Martin?"

"Of course. I just arrived from Lima. Remember? San Martin, himself, hung the Order of the Sun on me for my contributions to the patriot cause in Peru. His lover is one of my closest friends—Rosa Campuzano."

"My God! Why didn't you tell me before?" Bolivar rolled over and draped his arm across her naked body. "Darling, I can't take you to Guayaquil, but please, tell me everything about San Martin. Everything!"

"Why can you not take me?"

"Because, my angel, you are far too distracting. I must have my wits about me."

"And what if I insist?" Manuela asked.

"You're too smart to make that mistake," Bolivar replied. And he was right.

Chapter 34

General Jose de San Martin stood at the bow of a large Peruvian warship, the Macedonia. Under full sail, the ship resembled a sleek ocean-going yacht with shiny brass fittings and well-scrubbed teak decks; yet, when her gun ports swung open, she became the fiercest vessel on the water.

San Martin, a stout, rugged man of forty-five, stood alone until one of his staff officers, Rufino Guido, came on deck and joined him. "I've prepared everything for you, sir. Your notes are all in order."

San Martin nodded stiffly. "Thank you, Rufino." His bearing was decidedly military and somewhat aloof.

"You must be excited, sir. This will be an historic moment. The great San Martin and Bolivar coming face to face at Guayaquil.

The two most famous men on the continent."

"Yes. Yes. But remember, our meeting has a purpose." San Martin was always stern. His soldiers couldn't imagine him showing the slightest hint of excitement.

Rufino Guido nodded, speaking tentatively. "I know. But do you really think General Bolivar will help you liberate Peru."

"Certainly. You saw his letter. Now that he's freed his Great Colombia, nothing is more important to him than my success."

Rufino Guido looked doubtful. "I believe, sir, he said nothing was more important to him than the liberation of Peru. I'm told Bolivar is extremely egotistical. I'm not sure he'd join anybody as an equal."

"You mean he won't join *me*?"

Rufino elected to remain silent. His commander was extremely inflexible. It was never wise to argue with him.

The city-state of Guayaquil, located on the Pacific coast below Quito, was coveted by both Bolivar and San Martin. Some of its political factions wanted to annex Guayaquil to the Peru of San Martin and some still wanted to remain independent, although the majority desired union with Great Colombia. The Liberator settled the question decisively. Ordering San Martin's thousand men who fought at Pichincha to remain in Quito, he entered Guayaquil with a large segment of the Great Colombian Army at his back and was acclaimed president and absolute ruler of Guayaquil.

As people cheered, bells rang and the band played, Simon Bolivar pranced up the steps of the platform erected in the city square. The stage had been built so hastily that even now, workers were still hammering the last few nails.

Bolivar waved his arms to quiet the cheering crowds. "You were fighting among yourselves. You were faced with chaos. Nobody in Guayaquil knew what you wanted. But you are now part of Great Colombia!" The cheers and applause of the crowd became deafening. Bolivar raised his arms over his head and shouted, "You were threatened with anarchy. I have brought you salvation!"

* * *

As the Macedonia approached the port of Guayaquil the following day, Major Daniel O'Leary and General Simon Bolivar waited on the dock with the other officers hovering in the background. An honor guard of Colombian soldiers was mounting the flags of Great Colombia on every post and archway surrounding the piers. "General San Martin isn't going to be expecting this," observed O'Leary. "I think he's assuming he'll meet his own troops here and annex Guayaquil. How will he react to what you've done, sir? You have completely outmaneuvered him."

"What can he do?"

"They say he's one of the world's greatest military geniuses."

"It's true. But he's no statesman. Doesn't have the flair for it. Do you know how he styles himself? *The Protector.*" Bolivar grimaced. "After he freed Chile by defeating the Spanish so decisively in the Battle of Maipu, you'd have expected him to proclaim victory in ringing, memorable phrases. But do you know what he said? 'We have won the action completely.' That's all. Pretty dull stuff, in my opinion."

O'Leary grinned. "Certainly not the dramatic pronouncement we'd have expected from our Liberator. But he did free Argentina. And he crossed the High Andes to liberate Chile, and now he's taken Lima."

"So?" Bolivar answered defensively. "I liberated Colombia, Venezuela, Panama, Quito. And, don't forget, I crossed the Andes too!"

"Yes, of course. But I understand San Martin didn't lose nearly as many men in the crossing as we did."

Bolivar waved his hand impatiently. "He spent two years preparing his army for the move, for God's sake! I just went!"

"Yes, sir. I went with you, remember?"

Bolivar dismissed the subject impatiently. "Look, O'Leary," he said, pointing to the forward deck of the Macedonia, now preparing to dock. "I think I see him."

A stern-faced San Martin stared down at the activity on the dock and was unhappy with what he saw. It was clear now that General Simon Bolivar had thwarted his plans to annex Guayaquil to Peru. He could see the Liberator standing on the dock surrounded by his staff and the ranks of Colombian troops. Beyond them the flags of Great Colombia flew from every lamp post. Bolivar had beaten him to it and claimed Guayaquil for himself.

By the time he descended the gangplank, however, General San Martin had recovered his composure, and when the two greatest heroes of South America met, they embraced warmly. Each man then presented his staff officers and encouraged the group to mingle. The atmosphere was friendly, marked by laughter and good will. Amidst the pleasantries, Bolivar took San Martin by the arm. "We are both here to talk, I believe."

San Martin nodded.

"With your permission, I've prepared a house for our meeting —"

"The meeting will be between you and me only," interrupted San Martin. "Nobody else should be present. We must be able to speak frankly."

"But, of course. I wouldn't have it any other way."

For their meeting, Simon Bolivar had selected the largest house he could find in the Las Palmas section of Guayaquil. He had commandeered the finest French and Spanish furniture to appoint the room in which they would talk, including two intricately carved French chairs covered with silk cushions and a solid rosewood table.

When the two men entered that afternoon, San Martin looked around appreciatively. He glanced back at the large wooden doors. "Are they guarded," he asked. "I don't want anyone to disturb us."

"Yes. There are two sentries just outside." Bolivar sat informally on the large table with his left leg dangling to the floor. He smiled disarmingly and said, "I heard a funny story the other day. It seems a hussar met this ballerina—"

San Martin broke in at once. "I had intended to discuss the matter of Guayaquil. But I've changed my mind. The miserable inhabitants of this place misled me. They didn't tell me Guayaquil was already part of Great Colombia."

"It happened recently. You see, yesterday was July 24th, my thirty-ninth birthday. So, it was only natural that —"

"Let's not discuss it. We have more important things to talk about."

Bolivar shrugged, appearing to defer to San Martin.

"As I wrote you, I need your assistance. I cannot pacify Peru without the help of your Great Colombian troops."

"Yes, of course. Let me see, now. You sent me a division of a thousand men to help me take Quito—they fought bravely at Pichincha, by the way. Superb troops. So I'll tell you what. I'll send you a division of four battalions of 250 men each."

"You're joking! Come, sir. Let's be serious."

Bolivar spread out his hands. "Seems fair to me, sir. You sent me a thousand men. I send you a thousand—"

San Martin banged his fist on the table. "Stop being ridiculous! You know as well as I do that I cannot possibly defeat the Spanish without your entire Great Colombian Army."

Bolivar stood up and tried to look genuinely sad. "I'm sorry, sir, but I just can't spare the whole army right now. I need my troops here in Colombia. We still have a lot of Royalist diehards around. Spain is threatening us. Pasto's restless."

San Martin betrayed impatience, even desperation. "You can lead the Colombian Army into Peru. You would be in command of all Colombian forces."

"Oh, no. I can't do that." Bolivar began pacing the room, the heels of his boots clicking against the dark, hardwood floor. "You see, I'm still President of Colombia. I can't leave the country without the permission of the Congress."

"They'll give you permission. They will do anything you tell them to do."

Bolivar appeared doubtful.

San Martin sighed then spoke in a whisper. "I'll serve under you, sir."

Bolivar was momentarily stunned by the naked statement but recovered quickly. "No, I wouldn't feel comfortable with that. Besides, your army wouldn't let you. They'd revolt."

For the first time since he'd entered the room, General San Martin sat down. He shook his head in disbelief. "Is it possible that I stand in the way of Peru's liberation? Tell me, General, what is it you want? What can I offer you in exchange for your help?"

Bolivar's features betrayed no hint of his thoughts. "I hear you're having a few political problems in Peru." He took a few steps and returned to the table. When he spoke again, his voice was hard. "It would be senseless for you to take your army beyond Lima to force a battle with the Spanish in the mountains. They are too strong. They would beat the hell out of you. We both know that. But the Peruvians don't understand it. And your Argentines don't either. With all respect, sir, you've got a crisis on your hands."

The great Argentine slouched in his chair. His face reflected only sadness—the sadness of a man faced with another man's betrayal.

Off-handedly, almost languidly, Bolivar asked, "If you survive the present crisis and you're able to beat the Spanish, what are your plans for Peru? You will have to set up a government, you know. You will have to take over and rule."

"I'm no good at ruling. I never wanted power or fame. They don't mean anything to me. All I want is to leave a stable government behind me."

Bolivar nodded. "In that respect, we see eye to eye. Unless we can establish a stable government, there is no point to the independence we are both fighting so hard for."

San Martin sat up straighter. "Yes. I have to install a real ruler, somebody who knows how to govern, who was born for it and trained for it. I shall ask a European prince to be King of Peru. A good, solid monarchy is the only answer."

"No!" Bolivar moved away from the table, all pretense of languor gone. "Never!" His voice rose. "Not for Peru and not for Great Colombia. We're Americans. A European prince would be as out of place here as an elephant in the Andes. I shall fight against any monarchy. I shall fight it to the death."

San Martin was startled by Bolivar's sudden vehemence. "What would you have, then?"

"Liberty." Bolivar whispered the word.

"Liberty?" It was San Martin's turn to shout. He stood up and threw out his arms. His face reddened with anger. "Giving these people liberty would be like giving a two-year-old a box of razors."

"No. You misunderstand me. What I want is *liberty under law*. You have to grasp that. I shall never support a monarchy. Never. No king but perhaps a president for life. No nobility but perhaps a hereditary senate and an elected Congress."

"That's possible in the future. Right now, we need a crowned king to give us stability."

Bolivar cut the air with a swift slice of his hand. "No. No king. Not now. Not in the future. Not ever."

"Unless we install a king, there is no future."

The two men glared at each other. Both realized they had hit the point on which everything else depended, and neither was going to change his mind. After a moment, San Martin sat down again. "You think I want the crown for myself. Well, I don't. I wouldn't take it even if it was offered to me. I want to retire. I want to go back to Argentina."

Bolivar remained silent.

"Before I left Lima, I drew up my resignation as Protector and sealed it in an envelope. All I have to do is send a message telling them to open it. I don't even have to return to Peru." San Martin looked into Bolivar's eyes hoping to find some sign of encouragement.

There was none.

The next morning, Danny O'Leary appeared with his writing equipment, ready to take down the great Liberator's pronouncements on the results of his meeting with San Martin. But Bolivar waved him off. "Just say that General San Martin and I met, and I found him very soldierly," said the Liberator.

O'Leary hesitated, then said, "General San Martin says he misjudged you. He claims that you are not the man he thought you were. What did he mean by that?"

"No idea," replied Bolivar, dismissing O'Leary.

The meeting between Simon Bolivar and Jose de San Martin is considered one of the most important in the history of South

America. It left San Martin thoroughly disillusioned with Bolivar. He subsequently sailed for Europe with his daughter, where he retired into obscurity. His departure left the responsibility for completing the liberation of South America solely in the hands of Simon Bolivar, which was exactly what the great Liberator had intended.

Chapter 35

Peru was the treasure chest of South America and, arguably of the world. If the Spanish Crown could keep possession of Peru, they held the key to the continent. It was vital, and not only because of its innumerable riches. In the high mountains of Peru were two large, fully equipped Spanish armies, ready to attack in any direction—towards Lima, to Bogota, to Caracas, Panama, wherever they pleased. Together, these two Spanish armies constituted the greatest concentration of armed force on the continent. They had to be defeated or the fragile independence of South America would dissolve like an Andean mist and colonialism would triumph.

With the departure of San Martin, the Peruvians entreated Bolivar to come save their revolution. He was their only hope. Yet the great Liberator was now reluctant, and with good reason. A Spanish flotilla loaded with fresh troops sat off the coast of Venezuela. Although hoping the Spanish threat to Venezuela could be handled without him, Bolivar wanted to be close by in the event he was needed. The famous Juan Jose Rondon, whom the Spanish feared greatly, had died of an infection after receiving a minor wound in an insignificant skirmish. Without his cavalry leadership, the Spanish would be far bolder.

The most prominent, aristocratic, pro-Bolivarian family of Guayaquil was the Garaicoas who owned a large mansion, down the river from Guayaquil. Invited to stay at their home as long as he wanted, Bolivar moved in. The first evening just before dinner, the family was presented to the great Liberator. Mr. and Mrs. Garaicoa had three daughters, all considered to be beauties. The two older girls were quite proper and shy. They were unmarried only because the family considered themselves a cut above the other families of their class in Guayaquil and were waiting for the wars to end so they could betroth their daughters to more suitable young men from Quito or Lima. But the sixteen-year-old Joaquina was a natural-born flirt. Her skin was milky, her hair a shimmering mass of black onyx held in place by a large Spanish comb sprinkled with diamonds. Her face reflected a rare sensuousness coupled with intelligence and beauty. Her nose was long, which gave her a dignity she didn't possess. She was tall and stately, with a body men would be willing to die for. She seemed to realize this and moved in such a way as to accentuate every curve.

At dinner, she was bold. "I hear all your conquests are not on the battlefield, my Liberator," she said archly.

"My dear, my dear," intoned her mother, then turned to Bolivar. "Don't pay any attention to her, your Excellency."

Bolivar smiled. "I'd prefer you to call me Simon or, at the most, Liberator."

"May I call you Simon?" asked Joaquina.

Her mother sat bolt upright. "Certainly not, child!"

Bowing to Joaquina, Bolivar said, "Certainly, my dear." To her horrified mother, he said, "The youth is the future of our country, my kind friend. Besides, my mother married my father when she was fourteen and he was forty-six, and I'll bet you she called him by his first name."

Mrs. Garaicoa seemed mollified. Then, she had second thoughts. "Your father wasn't the president and Liberator, the leader of armies and of nations."

Bolivar smiled. "It would have been the same."

Joaquina turned to him and asked, "Would you marry a fourteen-year-old girl?"

Bolivar hesitated. At such moments he remembered his vow after the death of his wife. No, he knew he'd never marry anybody. He had found an angel and God had snatched her from him because he was unworthy of her. He would feel this way for the rest of his life. Thank goodness Manuela was already married, he thought. He could enjoy her company when they were together without ever having to worry about being pressured into something more.

He turned to Joaquina and said finally, "I shall never marry."

"But if you should, would you marry a fourteen-year-old?"

"I suppose so," replied Bolivar easily.

"I'm sixteen," said Joaquina.

The family and the Liberator roared with laughter.

"You're not very subtle, my dear," said her father.

The conversation ended as the ladies left the table.

After dinner, Bolivar and Mr. Garaicoa smoked their cigars on the balcony off the living room. "Unless you do something, Simon, the Spanish will be back in Lima and we'll all be slaughtered."

Bolivar nodded. "I know. But do you know there's a Spanish army off the coast of Venezuela waiting to land and take Caracas, then link up with their army in Peru? Do you realize that if they do that, our independence is finished, over forever?"

"Good God, no!"

Bolivar nodded.

"Do we have any chance? Any at all?"

Again, Bolivar nodded. "Yes, my friend, there is a possibility. If my plans work out, you'll learn about it quickly. I have made certain suggestions to the representatives of some powerful foreign states. They have been well-received, and if they bear fruit, I shall have covered my northern perimeter and be free to proceed to the south."

Garaicoa sighed with relief. "You are the hope of the continent."

"I'm also a very tired Liberator. With your permission, sir, I shall retire."

Instead of taking the grand staircase upstairs, Bolivar opted for the narrow wooden stairs outside which connected the ground floor porch with the upstairs balcony. The roof covered the stairs, giving camouflage and safety. Ever since the man named Pio had attempted to knife him in his sleep one evening in Jamaica, he made it a point to be circumspect in his movements.

At the top of the stairs he headed to his room. Gradually, he became aware that he was not alone. Then, he heard a rustling and discerned a lady standing beside the wall, away from the balcony. "Hello," he whispered. "Who are you?"

The lady inched towards him. "I'm Joaquina, of course," she whispered, giggling softly.

"Oh, how nice to see you." Bolivar was taken aback. The girl had obviously been waiting for him.

"I just love the fresh air, don't you?"

"Oh, yes," said Bolivar. "I especially enjoy it in the company of a pretty young lady."

"Like me?"

"Like you." Bolivar took her arm in his and walked her slowly to the balcony railing. "Look," he said. "Look at the night. Isn't it glorious?"

"It's absolutely dark. There's no moon. No stars, even. All I can see is the blackness of night."

Bolivar longed to reach out and hold her. She clearly wanted him.

But even Bolivar hesitated. This lovely creature was only sixteen and the daughter of his most ardent supporter. This is not only madness, he thought, but bad manners as well!

Putting his arm around her small waist, he said, "You're a very beautiful sixteen year old, too." He blew softly into her ear. He could feel her turning towards him. He reached for her face with both of his hands and kissed her on the mouth. She held him firmly around the waist and pressed her lower body against his. After they had kissed, she leaned her head on his shoulder and breathed hard into his ear. There was no misinterpreting her invitation as she gyrated against him. He pulled away, took her hand and led her towards his bedroom. She came willingly, breathlessly.

There was no more need for flirting. After one more long, hard kiss, both Simon and Joaquina quickly disrobed and entered his large bed. They made love all night. In the morning, Bolivar awoke, only slightly tired, to find that Joaquina had already departed. He smiled in the knowledge that he couldn't live without women—and that they couldn't live without him, either. He didn't know whether it was his fame or his personality that attracted them, and he didn't really care.

However, he did know he had to be cautious. Joaquina had been a virgin when she had climbed into bed with him last night. Her father and mother would be outraged if they knew of the affair. And Joaquina was such a madcap. She could easily boast of her conquest of the famous Liberator at the family breakfast table. He shrugged. He could deny the whole thing. She couldn't get pregnant by him, of that he was certain. But he decided to warn her at the earliest opportunity. Her father's support was too important to throw away on an adventure with his madcap daughter.

A week later, as Mr. Garaicoa and Bolivar were enjoying their after dinner cigar, the conversation turned to Joaquina. "I'm so happy to see that you and my daughter have become such good friends," Mr. Garaicoa began tentatively.

"She's a charming child," responded Bolivar, thinking all the while of the sixteen-year-old Joaquina lying in bed with him, sensuous and insistent. They had made love every night, all night, and she was beginning to exhaust the valiant Liberator.

"She'll be marrying soon," continued Garaicoa. "My wife thinks that despite your remarks the other evening you might be an interested suitor."

Bolivar smiled. "Oh, we're just friends. Most platonic, you know. She's a young girl, and I'm a well-known general. Hero-worship, I suspect."

In bed at night, he called her his "Gloriosa," the glorious one. One night, she explained to Bolivar she couldn't make love with him for a few days. He sighed with relief and told her he understood and not to worry. "Oh, no, my Simon. I cannot deprive you. Wait." In a moment she returned with her older sister, who entered the room shyly. Joaquina quickly stripped her naked and led her to

Bolivar's bed. Timidly, the sister entered, but soon proved to be as passionate as Joaquina, who watched the love making most of the night. Every once in a while she would giggle, or ask her sister if she didn't think that what she was doing was the most fun in the whole world.

The following night, Joaquina brought her other sister to Bolivar's bed and thoroughly enjoyed the entertainment. Afterwards, "La Gloriosa" refused to share the great Liberator and joined him in bed every evening, to their vast mutual enjoyment.

Chapter 36

Despite his strenuous evenings, the Liberator continued his intricate negotiations, to which he had alluded to the elder Garaicoa. He sat at his desk studying a sheet of paper with great interest. General Sucre knocked perfunctorily at the door, then entered Bolivar's office.

"You seem to be getting thinner, sir," said Sucre.

Bolivar shrugged. "I've been working hard."

"I've heard rumors," said Sucre. "You should know that everybody in Guayaquil except, of course, the older Garaicoas, know what's going on. Tell me, Simon, is it true? All three?"

Bolivar nodded sheepishly and smiled. He didn't delude himself.

"There are other rumors," continued Sucre. "These have nothing to do with the Garaicoa's bedrooms."

"They're true, Antonio Jose. The United States will not only recognize us. Their president, Mr. Monroe, will issue a declaration that no foreign power may establish colonies in America."

"What about the colonies the Spaniards already possess?"

"They may keep them—if they're able. But once we free them, they stay free. Spain cannot reclaim them."

"With due respect, sir, the United States is not a world power. Can they enforce Mr. Monroe's doctrine?"

"Perhaps not." Bolivar smiled. "But the guns of the British fleet can. And they will! Who do you think urged the Americans in the north to sign this thing? Our British friends, of course."

"You contrived the whole thing?"

Bolivar answered slowly. "In a way. But I was helped immeasurably by the fact that the United States are convinced the Russians mean to establish a permanent colony on the pacific north coast of America."

"On the other hand, sir, it's you who has the most to gain from President Monroe's doctrine," ventured Sucre.

"Without such a policy, do you realize what could happen, Antonio Jose? The Spanish could send troops over here every five years or so to try to take back their empire. We'd be fighting them off forever. How would you like that, my friend?"

"Not much. But, sir, please tell me. How in the world did you manage it?"

"That's my little secret. All I will tell you is that I kept my English contacts open. We are in close communication. But when I say secret, Antonio Jose, I mean *secret.* Suffice it to say that the British don't want the Spanish back here either. They want to trade with us. And the Americans want independent, republican governments in South America—not the Spanish Empire. Actually, they were about to issue a joint proclamation with England, but then, the American Secretary of State, John Quincy Adams, decided it should be completely American."

Sucre remained silent for a moment. "With the British fleet protecting our northern flank . . ." He was thinking. Now, he smiled. "We'll be going south."

Bolivar nodded once more.

Sucre threw his hat into the air. He slapped the Liberator on the back and shouted, "God be praised!"

The two men embraced happily.

* * *

Santander and other prominent Colombians wanted no part of an expedition to Peru, which they considered a foreign country. They believed it had been decided by all that Great Colombia would consist of Colombia, Venezuela, Panama and Ecuador, but the Liberator prevailed. He explained the danger of the Spanish armies in the Andes and the Colombian Congress voted him the express power to undertake the liberation of Peru while remaining president of Great Colombia. The Liberator was preparing to make the journey to Peru, when O'Leary came into the headquarters and cleared his throat. Looking up, Bolivar said, "Oh, good, Danny. You can help me load my papers."

"Uh, sir," said O'Leary, "maybe you should delay a bit."

"Why?"

"Well, sir, you know how news travels in these parts."

"Yes, of course. You think the Spanish are laying an ambush for me?"

O'Leary shook his head. "No, sir, but apparently Manuela has heard talk about you and Señorita Garaicoa, and rumor has it she's sore as a gumboil."

Bolivar shook his head pensively.

"She's also heard a ridiculous story about you and all three Garaicoa sisters, too. Nonsense, of course, but there it is."

Bolivar smiled. "Then, Danny, I'd better get away and go to Lima just as fast as I can. The more distance I can put between me and Manuela, the better—at least until she cools off a bit."

"That's the problem, sir. Manuela's already gone to Lima and she's waiting for you there—breathing fire, I hear."

Chapter 37

Simon Bolivar landed in Lima in September of 1823 hailed by every segment of the population as the savior of the country, the hope of the continent. The Peruvian Congress immediately named Bolivar Dictator of Peru with full power. The departure of San Martin had left two Spanish armies in the highlands ready to attack. The Peruvians were vulnerable, and they would have given Bolivar anything to bring him to their rescue. If the Spanish armies were able to descend on Lima without opposition, the reprisals would be ghastly. For his part, the great Liberator knew beyond any doubt that Peru was the key to South American independence. Without Peru and her riches of gold, silver and precious gems, Spain could not maintain her hold on the continent.

The Peruvians provided Bolivar with a mansion on the outskirts of Lima called The Mansion of La Magdalena. Much to his pleasure, La Magdalena turned out to be a beautiful estate with elegant columns, wide porches and colorful tiles. It was surrounded by flowing bushes, fig trees, small ponds and in the distance, the Pacific Ocean. This will be the perfect place to relax, Bolivar thought wearily. As always now, he was followed by a retinue of officers and politicians. He hoped they would leave him alone and let him settle in. The thought was intoxicating. Then, to the surprise of everyone, Manuela stepped out onto the front porch. It was well-known that Manuela and her husband, James Thorne, had lived in Lima for several years, and her moving in with Simon Bolivar and living openly with him was sure to cause a scandal.

She stood silently, beautifully attired in a full-length red and black velvet dress. She was smiling, Bolivar noticed, and fanning herself with an ornate black fan. And yet . . . he knew that smile. It boded him no good.

Reaching the door, Bolivar embraced her warmly and kissed her on each cheek. "It's so wonderful to see you again," he murmured, but there was no response.

In the marble-floored entrance hall, the Liberator told his entourage, "I'm tired. It's been a long day, and I must retire."

Entering his room, he removed his uniform coat and hung it on a peg. Servants scurried around, opening windows and smoothing down the bed.

"Out!" commanded a woman's voice from the doorway.

The servants dropped what they were doing and left hurriedly.

Manuela slammed the door behind the last of them and turned to Simon. Then, suddenly, she came at him. For an instant, Bolivar thought he was being attacked by a wildcat. Fingernails slashed, fists pummeled, teeth gnashed, and the words, "Your *Gloriosa,* is it? Your madcap lover, you call her? How dare you! How dare you flaunt your hideous indiscretions in front of the world? How dare you? Don't I mean anything to you anymore? You unfaithful son of a bitch! Three! All at the same time! I'll show you . . ."

"No!" he protested. "Not all at the same time!" Finally, Simon was able to catch Manuela's hands and stop the mayhem. She promptly kicked him in the shins, hard—and he fell to the floor. She stood above him like a heroic matador. He remained still, holding his shin in both hands.

Manuela let him lie there. She was angry—not because Bolivar had an affair; he would always have casual liaisons. She accepted that. And she herself was married. But this one could be serious. The possibility of Bolivar having a long-term relationship with another woman terrified Manuela. She couldn't bear the thought of losing him. She'd given up too much for him. And she knew she'd have trouble competing with a sixteen-year-old and, from what she'd heard, a sensuous, lovely sixteen-year-old at that. No. Let him lie there.

Bolivar opened one eye. "I am a stupid man. I was missing you so much . . ." He stopped and groaned in pain.

Seeing he was genuinely hurt and unable to get up, Manuela

knelt to cradle his head in her hands. "Oh, Simon," she asked quietly, "have I really hurt you, my darling man? Please understand, I'm a jealous woman. I have waited for you all my life and now that I have you, I do not want to share you." For the first time she was aware of the depth of her love for this man.

"You have killed me, my love," he said hoarsely. "But I forgive you even as I die."

Manuela pressed his face to her breast and stroked it.

Simon patted her back gently. "What could I do?" He croaked the words. "You know I need women. You weren't there."

Manuela was silent for a minute, thinking. He was right. He couldn't do without a woman, and she couldn't always be at his side. "Next time, darling, find an ugly old sow. In the dark we're all the same, they say."

"Oh, no, my dearest. You are in a class by yourself. Had you been there, I would never have gotten into that predicament."

"You really call her *your Gloriosa*, do you?" Manuela spoke softly this time.

"I couldn't remember her name. As you say, they're all alike in the dark."

"Madcap lover, is she?"

"Mad, perhaps."

"You liar! You're crazy about her."

"No. No. No. You don't understand. I was staying in her house. Parental supervision was lax. I was—I was *tricked* into the affair. They thought I'd get her pregnant and she'd marry me. That's all there was to it. And, as you can see, I left her in Guayaquil. I flew to your side."

"Oh, you poor darling," cooed Manuela. The fact that he'd fought to win her back reassured Manuela enormously. Otherwise, he'd have shrugged her off. This Garaicoa girl was just a passing fancy.

"But now you've hurt yourself, my dearest man. Let me put you in bed."

He nodded.

Despite his wounds, they remained in bed together for the next two days. Their reconciliation was sweet, indeed.

Chapter 38

Simon Bolivar encountered a morass in Peru. There were too many factions, too many rivalries, too many treasons. One Peruvian army mutinied. Three others simply scattered to the four winds. Without fighting a battle, they simply ceased to exist. Argentina was independent and secure, so the Argentine troops in Peru went home. The Chilean reinforcements decided they didn't want to die in Peru, either, so they turned around and sailed back to Chile. These events left Bolivar in Peru without an army. He wrote to Santander requesting six thousand experienced Venezuelan and Colombian troops, especially some Venezuelan plains cavalry units. But it would take months before Bolivar could expect to get the troops.

On his way back to Lima after settling a difficult insurrection in northern Peru, Simon Bolivar became so desperately ill he had to be taken off his ship at a small coastal town called Pativilca, where he lay between life and death for over a week. Some said the illness was caused by the tuberculosis; others said it was due to his strenuous love life. Most probably, it was a combination of the two, compounded by the almost insurmountable difficulties he encountered in Peru, problems so overwhelming that they sapped all the energy from his frail body.

In January of 1824, Joaquin Mosquera, the Colombian ambassador to Peru, a friend of both Bolivar and Santander, was returning to Bogota to report to the government on the 'Peruvian situation.' The only transportation between the two countries was by sea, and, en route, Mosquera, who had heard of Bolivar's indisposition, ordered his vessel to put into Pativilca. The boat docked with difficulty. The wooden pier was disintegrating; the men who handled the mooring lines were inexperienced. There was a lot of shouting between the ship and the shore before the vessel was made fast alongside. Colombian soldiers stood on the dock, watching the procedure while taking no part in it. They seemed suspicious. Although they appeared casual, they were on guard as if expecting some sort of a trick or treasonous act.

As soon as a rude plank was in place between the ship and the pier, Mosquera strode across it. He was a large man, portly without being obese, florid of face, yet handsome in a solid way. He walked with an energy that belied his girth. On reaching shore, he ignored the Peruvian officials and approached the nearest Colombian officer he could find. "I am Joaquin Mosquera, the Colombian Ambassador to Peru, and I have come to see the Liberator."

The man hesitated a moment, then nodded. "Please follow me, sir."

After a brisk but short walk, the two men came to a ramshackle beach side house that was a little bit larger than the other structures of the town, but in just as bad repair. The paint was peeling and the wood rotting. The officer opened the weathered wooden gate and motioned to Don Joaquin to enter. Inside the gate, two Colombian soldiers nodded at them and let them pass into the weed-filled garden. On a rickety old bench sat an emaciated man, his eyes cast down, his head wrapped in a white bandanna. Mosquera gasped and glanced inquiringly at the officer, who confirmed, "The Liberator, sir."

"He's dying." Mosquera's face betrayed his dismay. "The man's dying. He's almost dead."

The officer remained silent. Mosquera walked toward the figure on the bench, his pace unsteady. As he came closer, he saw a gaunt face deeply lined from pain and sorrow, arms and legs that resembled dried twigs off a dead tree, a wisp of hair protruding from underneath the bandanna. Mosquera, the tears now streaming down his cheeks, reached the bench and knelt in front of the Liberator. Bolivar looked up, saw him and smiled, extending a hand for Mosquera to shake. Holding Bolivar's hand, Mosquera found his voice. "This is terrible. What do you intend to do now, my general?"

Bolivar's hollow eyes lit up. Without hesitation, he replied in a firm voice, "Triumph!"

Before Mosquera left Pativilca, Simon Bolivar confided to him, "The only thing that keeps me going is that I can now count on something which I never had in my previous wars of liberation: *Fortress Colombia*. On her I can lean for support. From her I receive my strength. When I told you that I would triumph, it was not an empty boast. And I'll tell you why: In three months I shall have an army for the attack. With that army, I'll climb the high mountains and defeat the Spanish. Tell them that in Bogota. Tell them how you found me—fighting with bare hands and broken weapons to achieve the freedom of Peru and the safety of Great Colombia."

Joaquin Mosquera leaned forward. "I shall go straight to Francisco Santander. I shall insist the government send you everything you ask for. You can count on me, sir."

Two weeks later, a recovering Simon Bolivar kneeled beside Daniel O'Leary in the front row of the church in a small Peruvian town. Hundreds of candles burned steadily, their light reflecting off the massive gold figures on the solid silver altar. The church was filled to overflowing with Indians, who knelt in rapt worship. O'Leary seemed awed by the spectacle of so much gold and silver and lost in admiration of the sheer beauty of it. Bolivar was also absorbing everything, although his eyes kept coming back to the altar, and it seemed as if he were counting the gold candlesticks. But his face reflected not awe but avarice.

The Liberator seemed to recover as fast as he had fallen ill. A month later, Simon Bolivar was back in Northern Peru in the city of Trujillo, which was now his headquarters. The large room he occupied actually looked like a military command center. Maps covered the walls; soldiers stood at the entrance; officers came and went; secretaries sat writing dispatches, orders and directives for an army, even though there was no army, except in the mind of one man: Simon Bolivar. Still, he did have a few troops scattered over the landscape. The Colombian soldiers had begun to arrive. Marshal La Mar was assembling Peruvians, in addition to the men Bolivar had begun conscripting. His army would be assembled and trained quickly; it would be built up mainly around a nucleus of veteran Colombian soldiers—as soon as they arrived.

Beside Bolivar, in the middle of his headquarters, standing next to a table covered with papers, General Sucre was reporting to the Liberator. His face was sad and worried. "The Spanish are back in Lima."

Bolivar looked pensive. "Yes. After the Peruvians gave them Callao, it was inevitable that they would open the gates of Lima, too." He sighed. "It was, simply, unavoidable. Any more unpleasant news?"

Sucre shrugged. "Just before the Peruvian Congress resigned and pulled out of Lima, they named you Dictator of Peru again, with unrestricted powers." He hesitated. "Sir, may I make a suggestion? We all know Peru is a lost cause. Please don't accept the appointment."

Bolivar held up a hand. "My fate and the fate of my army is to save Peru or to die in Peru. There is no other road. You are free to go, of course. I don't expect you to fight for a cause you believe to be lost."

Sucre smiled wryly. "If the Spanish were to attack us right now, we would be completely defeated. They won't, though. Just the name, *Bolivar,* strikes terror into their hearts. Your reputation as a victorious general scares them to death. That's the only reason they haven't descended upon us before now. So I suppose I shall stay with you." Then, his smile broadened. "You are beaten. Everybody realizes you are beaten. But what nobody has forgotten is that when you're beaten you become the most dangerous enemy in the world."

"I knew you would stay. Is there anything else I should know?"

"Yes," said Sucre. "We have problems. That is, we have problems if you're determined to accept the post of dictator and the responsibility for the government."

Bolivar nodded curtly. "Go on."

"The troops we have are demanding to be paid."

Bolivar shrugged. "That seems reasonable. How long since they've been paid?"

"A year."

Bolivar jumped. "Well, dammit, pay them!"

"With what?"

The Liberator was briefly silent. "I shall raise the money. Have you ever been in a church here?"

"Of course. I go to Mass every Sunday and on Saint's Days and . . ."

"What do you see mostly in the churches of Peru? Even the smallest?"

Sucre looked puzzled. "Well, the usual things, Crucifixes, Virgins . . . And there's an awful lot of silver. And gold—candlesticks, entire altars, chalices . . ."

"I've requested the churches to donate their silver and gold to the cause—to pay my soldiers. I've intimated that if they don't, as Dictator of Peru, I have full powers to confiscate everything they've got . . ."

"But, sir! You're a good Catholic. You go to Mass. And you know how poor the Indians are. It's the beauty they see and feel in church that makes life bearable for them. Take that away, and they have nothing. Freedom won't mean anything to them, either. Their lot won't change. You hope to free Peru, but the Peru we're freeing is made up of the white people, the tradesmen, the aristocrats, the artisans and the land owners."

"I need money to pay my soldiers." He looked Sucre straight in the eye and saw dismay. He patted the younger man on the back and said, "Remember what I'm going to tell you: *War lives on despotism and is not waged with God's love.*"

After digesting Bolivar's remarks for a few moments, Sucre asked, "Will it be enough?"

"It will come close. But I'm counting on receiving funds from Santander. Great Colombia *has* to support us."

"Yes, sir."

"I'm giving you command of the allied army—under me, the Dictator of Peru."

Sucre cocked his head.

"You'll have the Peruvians under La Mar. But I'm keeping them on the coast for the time being, so they can't change their minds and go join the Royalists in the mountains. Cordoba will command the Colombians. They are the best troops you could ask for, and you know how brave young Cordoba is. Jacinto Lara will command the reserves—he's dependable and level-headed. You'll have General William Miller in charge of your cavalry. The man's brilliant. He's beaten the Spanish every time he's fought them."

Sucre smiled. "We owe San Martin something for leaving us *that* Englishman. I don't know how he does it, but he's got the toughest gauchos from the Argentine Pampas fighting alongside Chilean ranchers and Peruvian lancers. And they all worship him."

"Then, you're happy with the leaders I've picked for you?"

"Yes, of course," agreed Sucre.

"I'm expecting more troops from Colombia, but we're not going to wait. We'll start training and equipping the army right away. We'll have to go into the high mountains to fight the Spanish, and I'm going with supplies, warm clothes, food, ammunition, extra horses, mules . . ."

"You know we'll be outnumbered. We'll never be able to raise as many troops as the Spanish already have, unless we give ourselves more time, but I suppose that's out of the question."

"When haven't we been outnumbered? I'm more worried about their leaders. La Serna, the Spanish Viceroy, is clever. He had the good sense not to fight San Martin for Lima, but to pull out and bide his time in the mountains until the 'liberation frenzy,' as he calls it, subsides and he can take back all of Peru without effort. And Canterac is the best general the Spanish have. He'll be commanding their army."

Sucre was pensive. "Don't forget, too, they have a lot of native soldiers who are used to the high altitudes. They don't get sick or tired in the thin air the way our men do. And the Spanish armies have been operating all over these mountains for generations."

"I would have chosen any other place to fight—if I could have. Unfortunately, we don't have any choice. The only thing in our favor is that the Spanish are not expecting us to even try to climb those mountains. They think the very idea of such a venture will daunt us sufficiently to prevent us from coming." The Liberator smiled.

Chapter 39

Joaquin Mosquera, true to his word, met with Colombia's vice president in Santander's office at the palace, a tidy, well-ordered room without frills or luxuries. There were no curtains or pictures on the wall, no ornaments in the corners. And the furniture consisted of nothing more than one desk and eight stiff-backed chairs.

After a few perfunctory pleasantries, Mosquera came right to the point. "The Liberator needs money, men and supplies, everything you've got, General. You must know that."

Santander paced behind his desk, his face taut with anger.

"May I assume," Mosquera continued, "that you are in accordance with our Liberator's goals?"

Santander turned sharply and snapped, "There is no law which empowers me to assist Peru!"

Mosquera leaned over Santander's desk and said quietly, "If you hide behind the letter of the constitution, you are right—there is no such law. But surely, General, for the sake of freedom, you will rise above the law—"

"Our people have been bled white," yelled Santander.

"I know," Mosquera answered. "I can see it in the streets. They, too, are being asked to sacrifice."

"Not asked," Santander replied. "Forced!"

At that point a secretary, in civilian clothes but with a military bearing, entered the room and handed a document to Santander to review. After reading it, Santander said, "Listen to what I'm writing the Liberator. 'Either there are laws, or there are not. If there are no laws, why do we deceive the people with illusions? If there are laws, they must be strictly kept and obeyed.'"

Santander looked defiantly at Mosquera. "Colombia's inner security depends on the systematic growth of her constitutional life—which I shall defend with all my heart!"

"You are right, of course. But never forget that Colombia's external security depends on the independence of Peru. And Bolivar will fight for that with all his heart."

Santander laughed bitterly. "If I hadn't learned that by now, I'd be a very stupid man, wouldn't I, Mosquera?"

Despite his grumbling, Santander was coming through. Troopships were on the way from Panama and Guayaquil, loaded with Great Colombian soldiers, together with their battle equipment, arms, ammunition, supplies, extra uniforms for the Peruvians Bolivar was conscripting, and, in fact, everything needed for the campaign. By courier, he sent money to pay the soldiers and obtain materials locally. It was a strain on the population. It was a strain on Santander. But it had to be done. No one dared defy the Liberator.

Bolivar lost no time in organizing the army. Sucre was always up in the mountains making preparations for the trip. He had decided three columns could climb three times as fast as one could. He readied sites for spending the night. He cached food and extra blankets in these places. By the time he was ready, Bolivar was ready.

The columns of soldiers wound up the high mountains like long, multi-legged caterpillars. In places, there was space for only one man to pass at a time. Simon Bolivar led his files ever upwards. In the distance, on its way up a different mountain, crept another long column of men, who could be seen when the clouds blew away for a moment or two. A trumpet could be heard in the distance. Then a scream, as a man lost his footing and plummeted to eternity. When that happened, the other soldiers froze, afraid to budge. An officer had to shout, "March!" before they would move forward again. Horses neighed and mules whinnied. Officers from the lower reaches yelled to find out if they were on the right road; horns and trumpets from the upper levels told them they were. At twilight, the columns halted at more open locations chosen carefully by General Sucre in the months before. Blankets were laid out, tents pitched. Then came the smiling Indians scrambling up the mountains with the provisions for the next few days.

Bolivar warmed his hands by a fire. A young major came to join him. "We're at fifteen thousand feet, sir."

Bolivar nodded. "How are the other two columns doing?"

"Fine. General Sucre laid everything out perfectly. I think you were right to have your three divisions take three separate routes, sir."

"I shall feel a lot better when we've regrouped, and the army's all together again. I'm taking a chance by separating like this, but at least the enemy won't be able to destroy us all if he hits us on the march.

"It's the roughest march I've ever made. I never in my life thought I'd be clinging to the side of a cliff somewhere above the clouds. When I look *down,* I see high flying birds. And it scares the hell out of me."

At Pasco, a mining town high in the Andes, Simon Bolivar sat erect on his stallion and reviewed his troops as they proudly marched past. Beside him, also mounted, were General William Miller and General Sucre. As the army paraded in front of them, the distinguished-looking Miller turned to Bolivar. "Look at them, sir. These troops could hold a parade in St. James Park that would be the delight of London."

Sucre leaned over his horse's neck so Miller could hear him. "It's the finest army in America. The Liberator should look on it with great satisfaction. He created it."

Miller's brow furrowed. "He could have the finest army in

America, as you say, but until he got it over the mountains and in position to fight, it wouldn't be worth a farthing. I wonder if the enemy knows he's done that?"

The soldiers, having finished their parade, now stood at attention in arrow-straight rows, the cavalry to the right of the infantry. Bolivar rode forward and addressed them. "Men, you are about to complete the greatest task that Heaven ever assigned mortals: that of saving an entire world from slavery. The enemies you are about to crush brag of their triumphs of twelve years. They are worthy to measure their arms with yours, which have shone in a thousand battles. Soldiers, Peru and all America expects peace at your hands . . . peace, the daughter of victory. Even a liberal Europe smiles with pleasure upon you, for the freedom of the New World is the hope of the universe."

Chapter 40

As he rode along the shoreline of a shining, crystal-clear lake on a plateau of the Andes, Bolivar breathed the thin, pure air and surveyed the beauty around him. The majesty of the snow-covered peaks awed him.

Riding ahead of the cavalry with General William Miller, Bolivar checked his horse. He pointed. Miller shaded his eyes against the glare of the sun reflecting off the snow caps. "There they are," Bolivar said casually. "The Spanish Army has come out. They probably heard rumors of republican patrols operating in the area, and Marshal Canterac wants to investigate them." He grinned. "I'll bet he didn't expect to find our entire army here."

Miller had trained and fought with his cavalry all the way from Argentina to Lima and stayed on after San Martin left. Tall, his once handsome face now battle-scarred, Miller was an aristocrat to his fingertips. He now led the Peruvian Cavalry and fought for Bolivar.

Just behind them rode Colonel Laurencio Silva and Colonel Carvajal. Silva was a Venezuelan, tough and brave, with a keen sense of humor. He was observing the Spanish cavalry closely. Riding up to Bolivar and Miller he said, "Your plan won't work now. I know you wanted to attack the enemy from the rear. But look. Canterac's started to move his squadrons. See, he's drawing back."

"You're right!" Bolivar manifested his excitement for the first time. "There is no time to lose. Order your men forward at the gallop."

Bugles blared; metal clanked. Units that had been walking their horses, mounted quickly, as others handed them their lances. Led by Simon Bolivar, William Miller, and Laurencio Silva, the Argentinean and Venezuelan horsemen plunged forward towards the enemy. The Spanish continued to withdraw in an orderly fashion. Bolivar ordered his cavalry to slow to an easier canter. "The enemy commanders know this terrain like the palms of their hands. We have to be careful. They know they'll have to fight, but *they* will pick the place."

Miller and Silva nodded.

At five o'clock that afternoon on the Plain of Junin, the Spanish cavalry suddenly turned and charged. Their hoofbeats throbbed in the cold mountain air; trumpets echoed off the hills. The patriot cavalry drew itself up into battle formation just in time to receive the first shock of the Spanish attack. The clash of steel on steel was horrendous. The sheer weight of the Spanish assault carried their horsemen into the patriot's center and left flank. Sabers flashed, spears sung in the air and lances thrust forward with such power they penetrated enemy bodies up to their shafts. Bugles sounded and horses neighed. But since sabers and lances were the cavalry's weapons, and since no infantry or artillery units took part in the battle, not a single shot was fired. It could have been a medieval combat, fought before the invention of gunpowder. Bolivar, in the thick of the battle, rode from squadron to squadron, rallying his men, slashing and parrying with his saber. Yet the Spanish were carrying the day. They had penetrated deeply into the patriot lines and their victory seemed assured.

It was at this point that the Peruvian Hussar Regiment brilliantly led by General Miller struck ferociously. The gallant Miller, probably the best cavalry commander in America, if not in the world, and the two Venezuelan colonels, Silva and Carvajal, charged at the head of their troops, throwing themselves directly into the enemy cavalry with such force that the Spanish recoiled in confusion. The enemy tried to withdraw and reform, but Miller's horsemen pressed them relentlessly, preventing them from rallying. The withdrawal became a rout as the once invincible Royal Cavalry streamed from the plain—completely defeated.

The field of Junin was covered with dead and wounded men and riderless horses. The wounded cried. If their comrades—or their former enemies—didn't carry them to safety, they would die in the cold of the night. The Indians had already come down from the peaks to strip the dead of their uniforms. Vultures circled overhead. Night gripped the field of Junin in its icy fingers, and the howl of the Andean winds muted the screams of the wounded men and horses until, at last, the plain was silent.

Chapter 41

The bells rang wildly in Bogota. Fireworks brightened the night sky, and crowds cheered. Standing inside the palace, looking out the window at the scenes below, were General Francisco de Paula Santander and several members of the Congress. They were all frowning. The posters and newspapers lying on Santander's desk proclaimed in bold, black letters: "LIBERATOR WINS BRILLIANT VICTORY IN PERU!" and "SIMON

BOLIVAR TRIUMPHS OVER SPANISH IN PERU!" and, again, "BOLIVAR LEADS COLOMBIANS TO GLORY—DEFEATS ROYAL ARMY IN PERU!"

Turning to his adherents, his face sour, Santander said dolefully, "One more victory like this and the people will no longer call Bolivar their Liberator; they will call him their God. We must do something to stop this."

"But what?" asked a portly Congressman. "There is no way to stop him."

A sly look crossed Santander's face. He shook his head. "You're wrong. It will have to be done carefully and with skill, but I think I've contrived a way to prevent this kind of thing from happening again."

After his spectacular victory at Junin, Bolivar found himself in possession of half the highlands of Peru. In the mountain town of Huancayo, the Liberator sat in the town hall listening to the petitions of the people gathered there. The small building had a neat, tiled roof, large, wide-open windows, and walls washed white for the Liberator's visit. The floor, however, was packed dirt, and the only pieces of furniture were the table and chair Bolivar was using and several wooden benches, on one of which sat his escort of three officers. Everybody else stood. People walked in and out freely. A messenger entered and stood beside the large, open door without interrupting the proceedings. Bolivar was reading the names of the men he had appointed as magistrates, school teachers, clerks and administrators for the area. With each name, the assembled townspeople nodded approvingly.

Finally, the Liberator stood and stretched. The messenger took this as his cue and approached, saluted and handed Bolivar a sealed envelope. Still standing, Bolivar opened it with a small silver knife that was lying on the table. Before reading the letter, he glanced at the envelope; an official communication from the Government of Great Colombia in Bogota. He unfolded the letter and began to read. Immediately, he sucked in his breath and clenched his teeth. The officers on the bench all rose as one. It was impossible for them not to notice the Liberator had had a nasty shock. Bolivar drew another deep breath. "This meeting is recessed." His voice was barely audible.

As the people dispersed, Bolivar sat down and read the letter again. By now, his three officers stood in front of him, including young General Cordoba. Handsome and baby-faced, his black hair seemed to sparkle and his dark eyes flashed. He said anxiously, "Bogota has refused to send us more men."

The other two officers turned to him, astonished that anybody would dare interrupt the great Bolivar's thoughts.

Bolivar slowly shook his head. "No. In the last few months, they've granted me everything I asked for—men and horses, ships and ammunition. They're coming from Caracas to Panama, from Bogota to Guayaquil. They've given me more than I ever expected."

"Then what's happened?" ventured Cordoba.

Bolivar hesitated a moment before he answered, "The Congress in Bogota has revoked the Enabling Act. They've rescinded my powers to command the army."

"Not the army of Great Colombia. They can't do that."

Bolivar bowed his head. "They've passed a resolution. I'm forbidden to command troops on foreign soil."

"But why?" asked Colonel Laurencio Silva, his compatriot at Junin.

"I'm the President of Great Colombia. They say they need me and can't risk my getting killed and depriving the nation of my leadership when it is most needed."

"To hell with them!" Cordoba said angrily. "Tell them you'll do as you please and command the armies in Peru until you've beaten the damned Spanish. Then you'll ride into Bogota with your victorious divisions at your back and . . ."

"No. They've passed the law, and I must obey."

"It's not a law. It's an outrage." This was the sentiment of Colonel Carvajal, another hero of Junin who had been quiet until now.

"The army won't accept it." Cordoba pounded his fist on the table.

Bolivar stood up and raised his hand for silence. "Gentlemen, if I were to tell you I'm not hurt by this measure I would be lying to you. I have been deeply wounded by petty men, perhaps by the envy of one man."

"Santander," breathed Cordoba.

"But it's my pride that is hurt. My political position is as secure as it ever was. I am still President of Great Colombia and of Peru."

"Dictator of Peru," corrected Silva.

"I'm going to restore parliamentary rule to Peru just as soon as I can. In the meantime, I'm advising General Sucre I'm no longer in charge. He is to take full command of the army. I'm going to instruct him to tell the troops the news in such a way that neither their discipline nor routine suffers."

Then he looked directly into the face of each of the three officers. "And I want you to note and follow my example of obedience to the laws of the country. Do you understand me?"

Silva and Carvajal inclined their heads. Cordoba did not. Instead, he gave voice to his feelings. "Why? Every country you've liberated, you've liberated with your sword on the battlefield. You are a soldier. Civilians don't understand you and never will. And, sir, you don't understand them. Otherwise, you would never permit this."

"Enough." Bolivar waved his hand slowly in the air. "Even soldiers must obey the laws." But his words did not come forcefully.

After he'd dismissed the officers, Bolivar put his head in his hands. He felt as if he lived in a world of his own with dreams that no one understood. He looked to the future and imagined a place that was better than the one he lived in. But no one else, it seemed, could see it.

He would have to take them by the hand and show them, he thought. *First, we must win Independence.* Then, we institute a democracy. But he had to unify the people and persuade the local politicians to drop their petty jealousies!

When he looked up, he whispered to no one, "However sad my death may be, it will be happier than this life."

The long letter from the Liberator reached Sucre in his camp in the Andes, where he had halted his troops. He was waiting to discern the direction of the two Spanish Royal Armies so that he could out-maneuver them—or, if they united to attack him—elude or repel them.

He sat on a tree stump away from his men and reread Bolivar's letter. By now General Antonio Jose Sucre de Alcala was twenty-nine years old and considered the most brilliant general in the patriot army. Even so, he had repeatedly submitted his resignation from the army but each time the Liberator had rejected it. "You are too sensitive," Bolivar had said. "You listen to bad advice. You take offense when none is intended." Sucre knew Bolivar was right though he was still plagued with misgivings.

What did the Liberator mean, he wondered, by telling him the Congress had withdrawn the Enabling Act? Why had the Liberator placed him in command of all the patriot forces in Peru? Is he testing me? Or perhaps he feels the Spanish are too strong to be defeated; that I must be the one to shoulder the blame for losing the continent after he was so close to winning it. No, thought Sucre, Bolivar is not a spineless man. And he knows I can defeat any general in South America. My opponents and even my comrades-in-arms are naive; neophytes in strategy and tactics. For me, planning a battle is exhilarating. And most battles are won or lost before the first shot is fired.

Sucre knew he should feel confident and self-assured. But he didn't. His self-doubts kept eating on him. He didn't really like war. He thought to himself, I love the theory but detest the reality. Another cloud passed over Sucre's tortured face. The Liberator wants me to succeed him politically as well as militarily. I have no skills in governing. Nor desire. As president of Quito, everybody said I did a magnificent job, but they were only trying to curry favor with the Liberator for appointing me. Is he jealous of me? I can't help wondering.

I must ride to him. We must talk.

"What are you going to do now?" asked Sucre after Bolivar had reassured him of the truths contained in his letter, showed him the Congress's revocation of the Enabling Act. The candles flickered in the open, whitewashed *alcaldia,* the city hall high in the mountains of Peru. It was cold and both men wore ponchos.

Bolivar smiled and put his arm around Sucre's shoulder. "Actually, I think I'll go back to Lima."

"Sir, you know perfectly well a small group of Spanish soldiers have returned to Lima from their fortress in Callao. It's garrisoned with Spanish troops. General Miller smartly got his cavalry out just ahead of them and, thank God, he brought Manuela with him, or they'd have shot her for sure—Miller took a tremendous risk to take the time to find her and bring her with him. But you were shrewd to order her to stay behind when we moved out into fighting positions."

Bolivar nodded, still smiling. "I know. But I also know the Spanish troops in Lima are a small contingent. Otherwise, they'd have slaughtered most of the citizens. They're waiting to see what happens here in the mountains, where things will be settled once and for all."

"You'll be taking a good part of my army if you're determined to re-take Lima."

Bolivar shook his head. "No, no, no. I am forbidden by those little men in Bogota to lead an army here in Peru."

Sucre looked perplexed.

"I shall take only an escort of a dozen lancers," continued Bolivar.

Lima was buzzing with excitement. The whispered words passed from house to house, "Simon Bolivar is on his way to Lima."

Bugles blared as the Spanish garrison poured out of their billets, jogged down the streets to the *Plaza de Armas*, formed orderly ranks and immediately began marching out of Lima towards the fortress of Callao.

A few citizens poked their heads out of their houses. They waited. Suddenly, there was the sound of horses's hooves beating rhythmically on the cobblestones. "The Liberator's coming."

The cry was echoed by the hundreds of people pouring into the streets. "The Liberator comes. The Liberator comes." And then the shouts came true. Sitting straight as a ramrod on his stallion,

171

gracefully waving his hat and smiling to the crowds, Simon Bolivar rode into Lima. Looking behind the hero, a man called to a friend, "But where's his army? He's only got a dozen or so hussars with him."

"He doesn't need an army. He's the Liberator."

By now the bells were ringing in every church in Lima. With shouts of joy and wild cheers, the throngs had taken Bolivar bodily from his horse and were carrying him on their shoulders to his former headquarters. The entire populace of the city turned out. Their cheers shook the wooden balconies of the houses.

Chapter 42

On the morning of December 9, 1824, General Antonio Jose Sucre surveyed the plains of Ayacucho and the Spanish army that faced him. Peruvian Marshal La Mar, the brilliant English cavalry commander Miller, Colombian generals Lara and Cordoba stood beside him. "Look at their formations." Sucre pointed at the enemy in front of them. "They think they have us trapped. They think we are completely exposed. They've been so busy instructing the Indians how to slit our throats after the battle that they haven't bothered to reconnoiter the ground. They don't know it yet, but their cavalry can't operate the way they want them to. Our flanks are protected by ravines."

His generals listened appreciatively.

"Of course, they have the hills, and they have artillery. And look at their angles and troop concentrations. They are going to attack our left flank—that's you, Marshal La Mar. They'll try to force you back, then hit our center so hard it will crack and fall back on itself, after which it will all be over. They will have beaten us badly."

La Mar wiped his brow. "So, what is the plan now?"

"I'm positioning Jacinto Lara to reinforce you on the left, so they can't push you back." Sucre's voice was exceptionally calm. "Then, before they can hit our center, our right, under Cordoba, will attack. Miller, in the center, will support Cordoba's attack. Good luck, gentlemen. I must address the men."

As he rode to the front of his troops, Sucre's aide commented, "The Royalists have about ten thousand men, as far as we can tell, sir."

"We have less than six."

"They have artillery, sir,"

"We have one cannon with a broken carriage."

Sucre halted in front of his army. In a loud clear voice he spoke to the men, "Soldiers! How you fight today will decide the fate of South America." The ranks responded with loud cheers. They cheered Sucre; they cheered Bolivar; and—as an afterthought—they cheered the republic.

Not many moments after Sucre ended his speech to his troops, the Spanish cavalry attacked La Mar. At once, his men fell back while their reinforcements moved into line just as Sucre had planned. Jose Maria Cordoba dismounted and faced his infantry. "You have your orders. We go forward."

"At what pace, sir?" called a captain from the ranks.

"At the pace of conquerors. Follow me!"

General Cordoba then turned to face the enemy. He could see them plainly in their prepared positions. He walked briskly towards the Spanish lines, his men following in ordered ranks. The advance of Cordoba's infantry was valiant, even heroic. The men marched without firing a shot, even though they were raked by withering Spanish fire. They never paused or hesitated. Every third step they shouted, "Bolivar!" Two steps, then, "Bolivar!" Two steps, "Bolivar!" Beside them trotted General Miller's two cavalry regiments. The Spanish continued pouring their fire into them but were unable to slow them down, much less stop them.

Within range of the Spanish positions, Cordoba, calmly ignoring the enemy, turned to face his troops. "Prepare to fire! Aim! Fire!" The devastating fusillade literally blasted the Spanish line backward. There was, however, no time to reload. "Fix bayonets!" yelled Cordoba, drawing his saber. "Charge!"

"Charge, cavalry!" echoed General Miller.

The enemy could not stand against the fury of Cordoba's infantry and Miller's cavalry. Bayonets flashed until they were red with blood; sabers slashed down and down again; horses reared and plunged. The Spanish increasingly gave way; the patriot advance became irreversible. Colonel Laurencio Silva was wounded twice as he rode at the head of his hussars until finally he had to drop back. The field surgeon quickly bandaged his wounds and told him to go to the rear. Silva refused, saying, "Doctor, if I'm going to die, I had better enjoy myself on the way!" With that, he remounted and plunged back into the thick of the battle. He survived and was promoted to the rank of general.

La Serna, seeing his line crumble, shouted, "Send in the reserves! Send in the reserves!"

To no avail. Cordoba had driven the enemy into their rear entrenchments, overrun the entrenchments, routed the Spanish reserves, captured the Spanish artillery and taken the hill that supported the Spanish positions. As he climbed to the top of the hill, panting for breath, his saber dripping red, young Jose Maria Cordoba came face to face with a group of gold-braided officers. He approached them uncertainly, motioning his men to stay back, not to push on with their bayonets. A tall, stately gentleman stepped forward. His face was ashen white; his lips were straight slits clenched tight. He was holding his left arm, from which blood poured. He took a deep breath. "I am Don Jose de La Serna, Viceroy of Peru."

Cordoba was also breathing hard. He bowed. "I am General Jose Maria de Cordoba. In the name of the republics of Great Colombia and Peru, I take you as my prisoner."

On the battlefield of Ayacucho, Viceroy La Serna, Count of the Andes, shook hands with General Antonio Jose Sucre. "You have been most generous in victory. I was in no position to offer anything but unconditional surrender, and you allowed me to do so with honor. Thank you." La Serna signed the capitulation. So did General Canterac. Sucre's generosity to General Canterac—providing him safe conduct and the necessary facilities to return to Spain—was a supremely noble gesture. After the battle of Cariaco seven years earlier, Canterac had executed Sucre's brother, Captain Francisco Sucre, in cold blood.

After La Serna, Canterac, and their staffs had departed, General Sucre's officers crowded around him. Sucre told them, "You may spread the word. La Serna has surrendered to me all of his army, all of Peru, all garrisons and military supplies. We have won. It is over. The continent is free!"

The Battle of Ayacucho was, indeed, one of the most decisive engagements in the history of the world. It liberated all of South America from Spain forever.

Chapter 43

Simon Bolivar was at his desk in Lima reading petitions and various requests from his Peruvian constituents when an aide appeared at the door. "Sir, there's an officer outside with a message from General Sucre. He's been riding for eight days, all the way from a place called Ayacucho."

Bolivar leapt up from his chair. "News from the battlefield!" he cried and dashed into the hall where the messenger stood, muddy and wet, holding a bundle of damp envelopes. "Go get a hot meal and a bath," Bolivar said to him. "When you're dressed and fed, come back so I can reward you suitably." Bolivar smiled at the tired man, clearly boosting his morale and making the arduous ride from the battlefield of Ayacucho well worth the hardship.

Still standing in the hall, Bolivar read the report from General Sucre and beamed. After giving an account of the battle, Sucre had written, "Above all, I am happy to have complied with your orders. This letter is badly written and the ideas are confused. But, in itself, it is worth something; it brings the news of a great victory and the liberation of Peru. As a reward, I ask only that you keep your friendship for me."

Bolivar looked up at the officers who had gathered around him. Slowly, he unbuttoned his military tunic, removed it, tossed it onto the floor and stamped on it once. "Thanks be to God. I shall never have to command an army again." After that dramatic gesture, he danced through the palace at Lima crying joyfully, "Victory! Victory! Victory at last!"

The Congress of Peru was assembled in Lima, filling both the floor and galleries. The room buzzed with anticipation until, at last, Bolivar appeared at the hall doors. Silence fell over the room like a silken blanket. Slowly he walked to the podium, a small man and aged considerably, but huge in stature and fame. Simon Bolivar was truly the hero of this land.

A congressman whispered to his colleague, "It's him. It's the Liberator!" Then, as quickly as the silence descended, tumultuous cheers rose from the crowd. Applause echoed through the chamber; feet stamped the floor. "Great Bolivar!" roared the throng. "Long live the Great Bolivar!"

On the dais, Simon Bolivar raised his hands, basking in the adulation. Fifteen minutes later, he called for quiet but the audience was still in a frenzy. Finally, after several more minutes of cheering, the room fell silent. Bolivar cleared his throat. "Peru is free!" he began. "The Spanish are beaten at last." The crowd roared their approval. Then, speaking slowly, Bolivar declared, "It is now time for me to fulfill the promise I made to you. As of this moment, I hereby abolish the dictatorship. Gentlemen, victory has decided your destiny. A year ago I told you that the day on which this parliament convened would be the day of my glory, the day on which my most ardent desires would be fulfilled. This is the day. Now—once and for all time—I resign my rule."

Stunned silence greeted his words. No one made a sound. At last a portly congressman rose to his feet. "We cannot accept your resignation, sir. Peru needs you." Again, the cheers were deafening.

Another man rose. "We refuse your resignation, sir!"

"But I have decided," Bolivar protested. "My goals have been achieved. My work is done."

An older man, obviously a leader in the legislature, stood up. "Great Bolivar, you have exercised the unrestricted power of the dictatorship of Peru. Now, you wish to return to us the authority we invested in you during our time of crisis. But sir, you never abused your power. You leave behind no bloodstained walls, no opponents in chains, not a single crime of authority to dishonor you. On the contrary, you have led us to victory. At last we have thrown off the yoke of tyranny and can hold our heads high among the free peoples of the world. This kind of dictatorship, we must support with all our hearts. I demand that the people and the Congress of Peru reinstate you in all your powers." His closing words were drowned in thunderous cheers.

Bolivar was flustered. "No. My administration has been only a campaign, a campaign to free Peru."

The crowd chanted in rhythm, "Bolivar forever! Bolivar forever!"

"Listen to me!" Bolivar silenced the crowd once again. "Take these words to your hearts. Never grant unlimited power to one man—no matter who he is, no matter what he promises. The danger is incalculable!"

A rosy-cheeked man stood up and spoke with a strong, amiable voice. "Great Bolivar, here in Peru, as you have no doubt observed, nobody gets along with anyone else." The crowd laughed in agreement. "But sir, everyone gets along with you." He paused and waited for the thunderous applause to subside. "You are the unifying presence that binds us. Do not desert us. Without you there will be nothing but anarchy, nothing but . . ."

Like the others before it, this speech was lost in a roar of approval. "Bolivar forever! Bolivar forever!"

* * *

The Congress of Peru declared Bolivar the father and savior of their country. They struck a medal in his honor and erected an equestrian statue of Bolivar on the main square of Lima. In Great Colombia, statues, honors and a platinum medal were all bestowed upon him. Still, Bolivar wanted no part of it.

In the palace at Lima, he told O'Leary, "They keep thrusting crowns at me. These countries all want me to be king! I'm revolted by the very idea!"

"Yes, sir," said O'Leary.

"Listen to this letter from Paez in Venezuela: 'Napoleon was called upon by the most famous men of the Revolution to save France. You should become the Bonaparte of South America because this country is not the country of Washington.'"

O'Leary seemed pensive. "Did you reply, sir?"

"Of course. I wrote him *'You tell me this is not the country of Washington, and I say to you, this country is not France, and I am not Napoleon.* The title of Liberator is beyond any reward offered to human pride.'" He glared at O'Leary as if defying him to make a rejoinder.

Taking up the challenge, O'Leary said, "If they can convince you to accept a crown, sir, you won't be able to resign and leave them."

Back in his quarters, Bolivar continued thinking. The people wanted him to stay and to govern them. Should he give in to them?

"You know, you're right, darling," said Manuela Sáenz. "We shall leave America and go to Europe." She was reclining on a hammock on the porch of the palace at Magdalena. "At last we will have time for each other. Surely we deserve that now." She looked forward to traveling with Simon, receiving the acclaim and the glory with her lover.

Simon sat back in a comfortable chair, gazing out to sea. He loved Manuela, at least carnally, and certainly she loved him. A mere glance from her filled him with desire. But did he want her enough?

One thing Bolivar was certain of: If he took Manuela to Europe to live, they would eventually kill each other! He would have to go alone.

Manuela got off the hammock and came to Bolivar's chair. Kneeling in front of him, she clasped his hands and said, "Simon, darling, we'll be together, alone. No more wars to fight, no more congresses or politicians, no plots, no husband. Nothing but you and me. I know we can never marry, but you will stay with me forever, won't you, my darling?"

Bolivar smiled faintly, then said, "You know I will."

"I can't tell you how thrilled I am."

He squeezed Manuela's hand but turned his face away. For the first time in his life, peace was within arm's reach. To all the world, his mission was over; his life was his own to savor.

And yet, he was, at this moment, overwhelmed with lassitude and gloom. Where is the fire I once felt in my blood? he wondered. Where is the spirit? Why can't I grab it, embrace it? And how can I live without it?

Involuntarily, tears welled in his eyes, fortunately unnoticed by Manuela.

Chapter 44

At this time, Pisa, on the Arno River in Italy, was the home of a small but influential English colony made up mostly of writers, poets and other intellectuals. One of them was the famous Lord Byron, who was still mourning the death by drowning of his friend, the poet Percy Bysshe Shelley, who also had been a member of the English ex-patriot community of Pisa.

One day, while Lord Byron sat on a dock thinking of ways to honor Shelley and, at the same time, admiring his schooner, *Boli-*

var, the Countess Guiccioli strolled by, twirling a colorful parasol. "You certainly love that boat, don't you, darling," she said. "I wonder if Bolivar knows you named it after him."

Dreamily Byron shrugged his shoulders. "He probably doesn't know I even exist. But he will."

"Don't be silly. Everybody's heard of the great Lord Byron. You may not be as famous as Bolivar, but you are certainly not an unknown."

Byron turned to her. "I'll be sailing soon."

She seemed disappointed. "So, you're still determined?"

"Yes. I'm determined. I shall go to South America and join the great Bolivar. I'm no longer content to be known as a famous poet. I intend to be recognized as an illustrious warrior as well. I shall ride at the side of the greatest hero of them all—Simon Bolivar—when he leads the last charge of the last battle to liberate a continent."

"I hope you survive. General Bolivar lives a charmed life, I hear. He's always in the thick of the battle with a thousand falling at his side and ten thousand at his right hand, and he never even gets scratched. You'll be at his side—one of the thousand."

Lord Byron laughed. "No. I lead a charmed life, too. I always have and always shall." He stood up and shaded his eyes. He saw a horseman galloping towards them.

The rider pulled up when he reached Byron and his companion. He touched the brim of his hat. He looked out at the *Bolivar.* "I bring news, my lord. I don't think you'll be sailing to South America."

"Oh?" said Byron.

"The great Bolivar's army has routed the Spanish Royal forces and has already freed all of South America."

"When?" asked Byron. "Where?"

"Several months ago, my lord. High in the Andes Mountains at a place called . . ." here he had to consult his notes, "Ayacucho, my lord. And, before that, he led the charge that routed the Spanish cavalry at a place called Junin."

"Damn!" exploded Byron. Then he turned to the Countess Guiccioli. "Well, I suppose that leaves Greece. At least there's still one country where I can gain fame as a soldier."

The countess made a face. "Darling, I could understand your going away to help liberate a whole continent, but just one country? Greece?"

Lord Byron nodded sadly, his hopes of glory dashed.

Chapter 45

A visit to the newly liberated territories was an obligation and a necessity. Leaving Manuela behind, Bolivar departed from Lima with O'Leary and a military escort, all mounted. High in the Andes, the Liberator rode with his retinue at his back and Colonel Daniel O'Leary at his side. As usual, advance parties had gone ahead to secure proper accommodations for the Liberator and his entourage. "This has been a triumphal procession," observed O'Leary. "Every place we come to, the people receive you as the king of the world. The Indians up here look on you as father, God, and protector."

"They look on any ruler that way."

"Yes, but the gifts. Gilded saddles. Crowns of diamonds. Ruby hilted swords. And all you do is give them away."

"Don't complain, Danny. Remember, you're wearing those golden spurs they presented me in Arequipa."

O'Leary flushed. "The Peruvian Congress voted you a present of a million pesos. And you turned it down."

"I told them to donate the money to rebuild Caracas."

A cry was heard up the path. "Make way! Make way!" O'Leary laid his hand on the hilt of his saber until he recognized the Colombian army officer trotting up to them. Then, he relaxed.

The man pulled up his horse and saluted the Liberator. He was out of breath and had to pause. "The National Assembly in Upper Peru is convened, sir."

"That's exactly what I wanted to hear. You've brought great news."

"Sir!" continued the officer, who wanted to keep the Liberator's attention. "They want your protection."

"They shall certainly have it."

"They want to name their country Bolivar."

These words silenced even the Great Liberator.

On the return ride to Lima, in every city and town, throngs gathered around the Liberator and his retinue. There were endless speeches and receptions, flowers and gifts. The only opportunity Bolivar had to talk to O'Leary privately was when they were in the saddle en route. Leaving a large town early one morning, he turned to O'Leary. "You know I've been receiving a lot of letters from Manuela lately?"

O'Leary waited. He realized the Liberator had something on his mind and wanted to talk to somebody. He was flattered. He worshiped Bolivar. For his part, the Liberator knew how O'Leary felt about him, that in the young Irishman's eyes he could do no wrong.

"Her husband, Thorne, has returned to Lima. He's demanded she move into his house."

"He has the right."

"He knows about Manuela and me, of course. Everybody does. He told her he would forgive her, that he still loves her and wants her back."

"Yes, sir."

"She can't stand him. She tells me he's dull and brutish. And he'll never give her a divorce."

O'Leary seemed shocked. "A divorce? Sir! There is no such thing here. You know that. It's impossible."

"I think she's going to leave him, though. It's just as well, because I'm going to take her away from him anyway. I can't live without her. I've written her and told her so."

"Yes, sir." O'Leary's tone was emphatic. "That's fine. But no divorce, sir. That would be immoral."

All the nations in Europe and America had recognized the republics founded by Simon Bolivar. Ambassadors were already arriving. The extravagant gifts continued to pour in. Back at the palace in Lima, while O'Leary was making an inventory of the treasures conferred on the Liberator and the charities to which he intended to donate them, Bolivar confided, "Of everything I've received, I'm proudest of these." He picked up two letters from his desk. "This is from the Marquis de Lafayette, himself, sending me sincere congratulations. And this, Danny, is what makes me proudest of all: George Washington's family has sent me a beautiful letter together with his most prized possession, the gold medal struck for him after the Battle of Yorktown. I shall wear it and no other for the rest of my life. Imagine, Danny! Washington, hand in hand with Lafayette—the crown of all human rewards!"

And he meant it. From that time on, despite the hundreds of medals and decorations bestowed on him, to the end of his life, Simon Bolivar wore only *one*—the Yorktown medal of George Washington.

Walking up a wide tree-lined road outside Lima as they often did, Danny O'Leary commented, "You should be pleased. They're already calling you the second George Washington." He paused, pensive. "You know, sir, you're probably the most illustrious man in the world right now."

Bolivar waved his hand impatiently.

"Please hear me out, sir," O'Leary was saying. "I'm not trying to flatter you. You're the Liberator of a continent, the creator of nations. You have received every honor the world can bestow." O'Leary hesitated. Bolivar had stopped beside a large tree and gave O'Leary his full attention. "You told me you intended to resign all your offices and titles and retire to Europe," O'Leary stood facing him. "I think that is what you should do, sir."

Bolivar thought before answering. "That would be the easiest course for me to follow."

"It's the only course." But O'Leary saw the doubts flitting across the Liberator's face like clouds across the sun on a windy day. "You'll be acclaimed wherever you go. There will be fetes and balls and cheering crowds. There will be parades and speeches and receptions by the crowned heads of the most powerful nations on earth. They'll shower you with presents. Political leaders will seek your advice. Beautiful women will throw themselves at you."

Bolivar smiled. "I would love nothing better. If my conscience would allow me, I'd leave tomorrow. But these countries are so new. They don't know which way to go. They need me, Danny. I created them, and I must stay to insure their future." He looked questioningly at O'Leary.

O'Leary was slow to answer. When he did, he picked his words carefully. "You made a speech once. You called yourself the 'Son of War.' And, sir, that's what you are. You are the most victorious, successful general in the world right now. With your sword, you have freed a continent five times larger than Europe. But now the wars of liberation are over. You've won. You've accomplished your goal of independence, a goal everybody considered impossible. It's time to leave."

"No, Danny. I can't go. Too much is at stake. I must unite these countries I have freed—like the United States. The key word is 'united.' We must unite. Otherwise, I foresee nothing but revolution, backwardness and disorder."

The two men resumed their walk down the dirt road. "You see, Danny, when we unite, we'll need a strong government, but how strong should the government be? We fought for liberty, but we need order, too. Liberty pursued to its logical conclusion will result in anarchy. Order, carried to its extreme, will produce tyranny. I want neither."

O'Leary answered hesitantly, "I understand you. But perhaps you are not the only person who can make laws for these new countries. You have achieved peace and independence. Now, perhaps the civilians can learn to govern. Go away and see what happens."

"The leader of the North Americans in their War of Independence didn't leave. He stayed and accepted the office of president.

He made it possible for them to become the great nation they are becoming."

Bolivar's expression changed, and he pounded his fist into his hand for emphasis. "I shall stay."

O'Leary's heart sank.

To himself, Bolivar added, "But I am not the god everybody thinks I am. I'm a man. A mere human being. Still, I must try."

And try he did. But his efforts were not to everybody's liking. The Peruvians adored him and carried out his wishes happily. But his own Colombian and British officers weren't convinced. In the courtyard of a former colonial palace, now a barracks, Rupert Hand approached General Jose Maria de Cordoba, the undisputed hero of Ayacucho, who was five years younger than he. "What news, sir?"

"The Liberator's staying on." Cordoba did not sound pleased. "And he's getting energetic and intellectual again, and messing everything up, as usual."

"Like what?"

"Schools. He's setting up all these damned schools!"

"He's always thought education was a good idea. So do I. In fact, if I weren't a soldier, I think I'd be a school teacher." Hand was sincere.

"You don't understand. Our Liberator has decreed that the children of the poor shall receive free education!" Cordoba was really upset.

Hand looked perplexed. "Schools cost money. Teachers have to be paid . . ."

"He's decreed the government will pay for educating the children of the poor. How do you like that?"

Hand seemed pensive. He nodded, thoughtfully. "Yes," he said. "But we're not ready for that sort of thing yet. The man's a radical. This is going to be received as badly as . . ."

". . .As when he freed the slaves?"

"Yes, I'm afraid so." Rupert Hand was upset. "Our Liberator is too far ahead of his time."

* * *

Colonel Daniel O'Leary sat writing a report. His brow wrinkled as he sought the right word. Colonel William Fergusson walked into the small office, which adjoined the Liberator's. "I say," began Fergusson.

O'Leary looked up and smiled. Like everybody else, he couldn't help liking Fergusson. "What can I do for you today, sir?" Even though he held the same rank as Fergusson, he still called him "sir." It was automatic and conveyed the respect of a younger man for an older one.

"Is it true we can't go hunting llamas anymore?"

O'Leary nodded. "The Indians live off the llamas, sir. We've been killing them for meat and fur blankets until there's a chance they might be exterminated, which would make it hard for the Indians to survive. That's the reason the Liberator issued those decrees protecting llamas by law . . ."

"Look, Danny, I can understand him not letting us shoot up the Indians, but the llamas are animals. I don't understand it. Nobody else in the world does anything like that. I hear that in the new territories of the United States they can kill all the damned buffalo they want to. And there are still plenty of them."

Manuela Sáenz lay seductively in her favorite hammock on the veranda of Bolivar's villa at Magdalena, outside of Lima. She heard hoofbeats and horses pulling up in the courtyard. She smiled. Simon was returning home after an active day. A few minutes later, Simon Bolivar came through the door from the bedroom, leaned over and gave her a kiss, patting the bottom of the hammock at the same time. After the perfunctory caress, Manuela swung her legs to the floor but remained sitting on the hammock. "I hear you've been thinking up projects again, darling. Things that get everybody upset."

"Oh? Which ones get people upset?"

"They all do. But recently you've been talking about digging a canal through Panama again."

"Certainly. A canal through Panama means we'll be able to send ships from one coast of Colombia to the other." Bolivar looked pleased. "And, just incidently, it will unite the entire world commercially by joining the Atlantic and the Pacific Oceans."

Manuela coughed discreetly. "A lot of people, including some of your beloved bishops, feel that 'What God hath joined together, let no man put asunder.' I think you will have trouble convincing them on that one. And some of your other friends say it simply cannot be done. It's an impossible task."

"Don't worry. A canal will be built at Panama. It has to be done, and if I don't do it somebody else will."

The Liberator's word was law, and he suffered only one setback—and this was due entirely to his own misjudgment. Bolivar decided that the nations of the Western Hemisphere should form what he called an Organization of American States to arbitrate disputes among its members and unite against external aggression. As he put it, it was to be a league of good neighbors, and for this purpose he organized a Pan American Congress to meet in Panama. This was to be the permanent seat of the association, of which he wanted the United States to be a member. The United States agreed to send observers and every other nation was enthusiastic about the project. It was to be monumental—until Bolivar announced that he would not attend because he had decided the countries of the region had to develop their own leaders. Since all the heads of state and prominent men of the Western Hemisphere countries were making the arduous journey to Panama solely to meet and talk with the great Simon Bolivar, his announcement served to discourage the other great men of the continent from attending.

Therefore, the meeting was held without him, as he had decreed. The United States's delegates died on the way to Panama. Argentina didn't send an envoy. Neither did Chile. The Bolivian delegates arrived too late to take any part in the proceedings and Brazil, being a monarchy, didn't even answer the invitation.

To outside observers, it was now obvious that without the Liberator's tremendous prestige to hold them together, the new nations of the hemisphere would have already flown apart like a pocket watch hit by a hammer.

Chapter 46

In Bogota, General Santander sat at his desk in his still Spartan office. The only addition was an oil painting of himself, General Francisco de Paula Santander, in full uniform, a gift from the members of the Great Colombian Congress. The door to the office was open, and a young aide in civilian dress entered. His round, pink face wore a perplexed expression. He cleared his throat and said, "Sir, there's a man outside who says he's come to see Simoncito Bolivar, sir."

"*Simoncito* Bolivar? Who the hell would call our Liberator *that?* Some nut, I suppose? A crackpot."

"He comes from Europe, sir. Speaks French."

"An emissary of some kind, do you think?"

"No, sir. He's not dressed very well and he's old and stooped. Almost shabby, I'd say."

Santander was about to wave his hand to indicate to his aide the man should be sent away without another thought.

"His name, he says, is Doctor Simon Rodriguez, sir."

Santander leaned back in his chair thinking. "Simon Rodriguez," he said slowly. "Well, I'll be damned."

"You know him, then, sir?"

Santander shook his head. "No. I never met the man. But I know who he is. And so do you, if you'll use your head for something besides a hat-rack."

But the aide was too young to have heard of Simon Rodriguez.

"The tutor," said Santander. "The Liberator's tutor, you dolt."

"Oh," said the aide. "Yes, sir."

"Well, I'm damned curious. Show him in. I want to meet him."

"He's with his wife, sir."

"Fine. Then, show the couple in."

Almost before the words were out of the aide's mouth, an old man, grey-haired and balding, with broken teeth came through the door accompanied by a lovely black-haired girl, who didn't look much older than thirteen, despite her swollen, pregnant belly.

"Why do you keep me waiting, young man?" asked Rodriguez. "You must know who I am."

The aide stood mutely at attention, not knowing how to respond. General Santander came around his desk and shook the old man's hand. "Of course, I know who you are. Unfortunately, your old pupil is away. He's off adventuring in Peru, where he has no excuse to be."

"Oh?" responded Rodriguez. "Well, then, tell him I'm here, and I'm sure he'll return at once."

Santander nodded. "An excellent idea," he said. "Would you like to write to him to let him know? Or should I?"

"Who are you?"

"I am General Francisco de Paula Santander, Vice President of Great Colombia. Even though, as you can see, my office is poorly equipped, that is only because your pupil uses all of our money for his foreign adventures."

During the conversation, young Mrs. Rodriguez played with the items on Santander's desk. Standing on tiptoe, she reached out for an ink well, knocked it over, shrieked and ran to Rodriguez. "I didn't mean to do it," she wailed.

"It's all right, child," said Rodriguez, patting the top of her head.

"Look," said Santander, "I'll write and advise the Liberator you're here. You can stay in Bogota until I hear from him."

"Where shall we stay?" asked Rodriguez. It was then that Santander realized he was facing not one but two children.

After their meeting, General Santander decided it would be good politics to lodge Rodriguez and his wife in a suitable boarding house in Bogota while he advised Bolivar of his arrival and waited for a reply. By return post-rider he received a note from the

Liberator to send the Rodriguezs to Lima, that he was going to appoint his old teacher the Minister of Education in Bolivia. With his note to Santander was a letter addressed to Don Simon Rodriguez. It was not sealed. Bolivar obviously wanted Santander to read it, which he did.

But Santander was having problems in Colombia. Paez was doing as he pleased in Venezuela and refusing to acknowledge the authority of the central government in Bogota. General discontent was leading to small uprisings all over the country and Colombia's financial condition was dismal. The government was in trouble, and Santander wanted Bolivar to return to handle Paez and anybody else who refused to obey the constitution. However, he had to take care of the matter of Simon Rodriguez and his pregnant wife if he wanted to stay on the Liberator's good side.

Simon Rodriguez once more stood before the vice president of Great Colombia, General Francisco Santander. The vice president smiled and handed Rodriguez the letter. "The Liberator is upset with you for not letting him know you are here in Colombia," said Santander. "Please read this."

Rodriguez read aloud, "Oh, my teacher! My friend! My Robinson! To think that you are in Bogota and have neither written nor addressed me."

"You should have written him," Santander interrupted. "He wants you to join him in Peru. He has big plans for your future."

Rodriguez inclined his head and continued reading silently. After a minute or two, he looked up at Santander. "Have you read what he says to me?

"He says: *'You molded my heart for liberty, justice, greatness and beauty.* I have followed the path you laid out for me. You cannot imagine how deeply engraved upon my heart are the lessons you taught me. I have followed them as infallible guides. In short, you have seen my thoughts in print, my soul on paper, and you must have said to yourself: All this is mine; I sowed this seed. I watered this plant. I strengthened it when it was weak. Now it is vigorous, strong and productive. Behold, here are its fruits. They are mine. I shall enjoy the shade of its friendly branches, for that is my inalienable right and mine alone."

Simon Rodriguez looked up from the letter. A whimsical expression crossed his face. "He thinks I'm a gardener."

Santander roared with laughter. He had heard that Rodriguez was eccentric. Now he knew the man lived in another world.

Despite the Liberator's later efforts to salvage him—he actually did appoint him Minister of Education of Bolivia—Rodriguez failed miserably. He went into classrooms and pranced around naked to demonstrate the freedom of man as a natural being. That was too much for the conservative Bolivians to accept. Then, the old man, having deserted his girl-wife, impregnated another thirteen-year-old and married her. Requested to leave the country, Rodriguez and his new bride made their way to Peru and Simon Rodriguez ended his days teaching poor children for free and earning his living making and selling candles.

Chapter 47

The frigate, *U.S.S. United States,* dropped anchor in Peruvian waters in February of 1825. Sent to Peru as a goodwill gesture to celebrate the initiation of relations between the U.S. and the new Peruvian government of Bolivar, the warship was commanded by the illustrious Commodore Isaac Hull, a hero of the War of 1812.

It was Hull who commanded the *Constitution,* known as "Old Ironsides" in her famous battle with the *Guerrière,* the most important single victory in U.S. Naval annals. Besides being an acclaimed warrior, Commodore Hull was diplomatic enough to have been selected for this mission and famous enough to have had his portrait painted by Gilbert Stuart. With him he brought his wife and her sister, Jeanette Hart of Saybrook, Connecticut. Hull's first act was, quite correctly, to invite the Liberator on board for an official dinner. Bolivar gallantly suggested the 22nd of February, the anniversary of George Washington's birthday.

The dinner guests gathered in a small but lavishly decorated dining room aboard the vessel. Yellow velvet curtains covered the port holes and ornate silver candle holders lined the table which was long and narrow with clawed feet.

At the dinner, Commodore Hull raised his glass and toasted Simon Bolivar as the George Washington of South America, the Liberator of a continent, worthy of mention in the same breath as Washington, Lafayette and Jefferson.

Bolivar rose and toasted the United States and all her heroes. His speech was fluid, complimentary and well-received. However, when he made his most flattering remarks, he gazed into the eyes of the lovely Jeanette Hart, who was seated at the end of the table. Her brown eyes and dark ringlets, which flowed to her bare shoulders, had caught his eye at once. She seemed intelligent and poised. After dinner, the South Americans and North Americans mingled, and Bolivar lost no time in making his way to Jeanette. "I am so pleased to meet you and your distinguished family," he began in English.

Jeanette laughed flirtatiously and replied in perfect French, saying she understood he was more fluent in that language.

"I am delighted," responded the Liberator. He noted that Jeanette was wearing a Regency gown that accentuated her generous bosom.

"No, your Excellency, it is I who am delighted to meet the George Washington of South America. I am absolutely thrilled. You are the greatest man in the world and I can't believe I'm actually talking to you!"

"The George Washington of South America!" Bolivar beamed with delight. "I'm flattered beyond expression." Settling down to after-dinner flirting, he continued, "It was so thoughtful of your sister and brother-in-law to invite you on this voyage of good will. You add a touch of feminine elegance to the enterprise, and I am not ashamed to say it: you fascinate me."

Flustered, Jeanette began to fan herself vigorously. Her cheeks turned crimson. Taking a deep breath, she said, "Your Excellency, in the north, gentlemen do not say such things to ladies they have just met."

"Ah," said Bolivar, ignoring the fact she kept calling him "Your Excellency," "but in this country gentlemen speak more honestly. I cannot find words to describe your charm and beauty. It is as if God had dropped a priceless jewel upon this fortunate vessel. You shine and you sparkle."

"Please, sir. You're embarrassing me." Her fan was moving even more rapidly, although the evening was quite cool.

"I intend to reciprocate your brother-in-law's kindness and hold a small fete for him and his officers. You and your sister will attend, naturally."

"I shall be honored, Your Excellency." Jeanette curtsied prettily.

At that point they were interrupted by Commodore Hull, who took the Liberator by the arm and said gruffly, "Come. sir. You must meet the rest of my officers."

"And also chat with Mrs. Hull," responded Bolivar, smiling. He bowed to Jeanette and took his departure, leaving a breathless and overwhelmed young lady in his wake.

The Liberator's party at the palace outdid anything that could be contrived within the confines of a vessel of war. Several hundred people were dancing and flirting, the wine was flowing, and the dinner fare delicious. Bowing in front of Jeanette Hart, Simon Bolivar, President of Great Colombia and Absolute Dictator of Peru, asked if he could have the honor of the next waltz. "Certainly, sir," she responded, smiling.

Holding her in his arms as they danced, Bolivar could feel her breathing hard. Keeping her at arm's length, as was the custom, he could enjoy the wonderful view of her full body. They twirled with the rhythm of the music, their movements sensuous. During a break in the dancing, Bolivar beckoned to O'Leary. After presenting his young officer to Jeanette, he said, softly and in Spanish, "Danny, please escort the commodore and his lady to the veranda.

They seem much too interested in this lady and me, like a pair of dueñas."

Bowing, O'Leary made his departure and did as he was instructed.

By then, the music had recommenced and Bolivar, once again, took Jeanette in his arms. Only this time he danced her right off the floor and into his study, closing the door behind them.

"Where on earth have you two been?" thundered the commodore.

"Yes, my dear," echoed his wife. "Where have you been?"

"You are not to worry, my dear friend," said Bolivar. "I have been showing this lovely and charming young lady around the palace and the grounds. There is quite a bit to see, I assure you."

"Your Excellency, it is getting late, and I am afraid we shall have to return to the vessel."

"But of course, of course. I shall have the carriage brought around immediately. Unless you'd like to take the tour I just gave Jeanette . . ."

"No, no. We're ready to go."

Two days later, an invitation was received on board the *U.S.S. United States* requesting Miss Jeanette Hart to join a group of Peruvians and French residents for a picnic. The invitation was most proper and written on the stationery of the Liberator, himself. It could not be refused and certainly seemed innocent enough. "He's inviting you because you speak French," said Hull.

On shore, it came as no surprise to Jeanette that the other members of the party had all taken ill and would not be able to attend. The spot Simon picked was lovely and shaded with a brook running by. The driver of the carriage, Jose Palacios, immediately departed, and Jeanette and Simon were alone. "The perfect setting," said Simon, "for the perfect lady."

Jeanette cast her eyes down which was the proper response for a lady of the day to such an extravagant compliment.

Bolivar took her hand in his. She did not pull away. "These two days without you have been an eternity," said Simon. "I don't know how I was able to stand the anguish. I dreamed of you at night, and when I awoke I saw your face before me. Each night, I took you to my bed . . ."

"General Bolivar!"

"In my imagination, of course, and only out of longing for you." He squeezed her hand. Taking one of the silk pillows they had brought along to sit on, he said, "Here. Put this under your head."

"I think I'll sit for a while longer."

Bolivar smiled. "As you wish, my dear. I have something I wish to present you. It's something personal, and something I hope will be agreeable to you." He reached into his coat and retrieved a small jewel box. Extending his hand, he offered the box to Jeanette and she took it with a pleased but puzzled expression. Opening the box, she gasped and then smiled radiantly as she removed a gold chain. Attached to it was a miniature portrait of none other than the great Simon Bolivar.

"Is this for me? To keep, I mean."

"It is, indeed. And I do not have to tell you there is only one. This is it."

Jeanette threw her arms around his neck. Her ecstacy was transparent.

"Now," said Simon, "please allow me to place this pillow under your beautiful head."

Upon returning to the ship around sundown, Jeanette Hart went to her stateroom and sat down. Taking pen in hand, she began to write. By midnight, after several interruptions by her sister, she had composed a most beautiful poem in French.

The next morning the poem, sealed in an envelope, was delivered to the Head of State of Peru. Bolivar recognized the handwriting and opened it at once. The staff officers standing in front of his desk saw him smile. He finished reading the entire poem before turning back to the business of the day.

Later, Bolivar wrote to Miss Hart.

I would like, mademoiselle, to be able to answer you in a language worthy of the muses, and worthy of you, but I am, alas, only a soldier. Your charming verses are so flattering to me that I do not hesitate to find them more sweet than the singing lyre of Orpheus. O Wonder! Young beauty singing of a warrior. Only gratitude saves me from annihilation and gives me speech to interpret my admiration and attachment to you.

There were more poems and more "picnics." And, of course there were dinners and balls. Jeanette Hart and Simon Bolivar were inseparable. But Commodore Hull was becoming anxious for his young sister-in-law. To his wife, he confided, "I have heard from our Peruvian friends that the Liberator is not to be trusted with women."

"Oh?" said Mrs. Hull.

"They tell me he is a great libertine. We must keep a sharp eye on Jeanette. She seems to have caught his fancy."

"Yes, of course. But it will be hard. Remember, Isaac, he's the greatest hero in the world right now. What woman can resist him?"

"I know one who'd better."

"Besides, darling," said Mrs. Hull, "I've kept my ears open too, and he's bedding every lady in Lima every night of his life, so he can't possibly have any time left over for our dear, sweet Jeanette."

The commodore still looked uneasy.

Jeanette woke early, aware of the sensation of the frigate rolling in the open sea. She jumped out of her bunk, dressed hurriedly, and dashed up the companionway to the open deck. The sea breeze hit her in the face. We must be going to another Peruvian port, she thought. Looking desperately for her brother-in-law, she ran aft and found him walking the deck. "Where are we?" she asked, shouting against the wind.

Hull smiled. "At sea, my dear. We're going home."

"But, but, Isaac, we haven't said our farewells. We're being terribly rude. Your mission will be a disaster, a failure."

"No, my dear, your sister and I went ashore last night and bid our hospitable friends goodbye and thanked them for all of their kindnesses. We included you in our expressions of appreciation, so you're not to worry."

"Not to worry!" Jeanette flung herself against the ship's rail. Her shoulders heaved, as she sobbed.

After her return to the United States, though she had many suitors, Jeanette Hart never married. She kept Bolivar's faded letters for the rest of her life and, to her family's distress, she joined the Roman Catholic Church. What had been an intense, even sincere, romantic fling to Bolivar had a much deeper and more lasting significance for Jeanette. She never saw him again but she loved him until the day she died, still wearing his miniature around her neck.

Simon Bolivar walked slowly and sadly into the villa at Magdalena. It was a cool and cloudy afternoon with a light breeze rippling through the trees. His feelings were bitter-sweet.

Manuela reclined in a hammock nearby with a blanket draped across her legs. She was smoking a cigar, a look of smug satisfaction on her face. She sat up as Bolivar approached. "Why so glum? You look like you lost your last friend."

"Oh, nothing," he replied. "The Americans sailed last night. Commodore Hull and his lady came to bid me farewell."

"And Jeanette?" asked Manuela, grinning.

"No. She remained on board."

"What a pity, Simon. My heart bleeds for you."

Bolivar stared at her coldly. "Your sarcasm is not amusing. Jeanette was a lovely girl, that is all."

Manuela threw off the blanket and jumped up from the hammock, her rage barely contained. "That is all? You are a liar, Simon. An unfaithful, heartless, liar!"

Bolivar backed away, seeming somewhat more contrite. "How did you know?"

"Everybody in South America knows!" Manuela screamed. "You couldn't possibly have been more indiscreet."

"I think you are overstating things, my dear —"

"You are a fool, Bolivar. A fool for women." She paused and drew heavily on her cigar. Then, exhaling a cloud of smoke, she said, "And I am a fool to love you."

He looked up thoughtfully and muttered, "Perhaps."

"You do know why she didn't say goodbye to you, don't you?"

Bolivar shook his head.

"Because I found all those lovely poems she wrote you. In French no less! I saw her the other afternoon at a salon, and I told her it was time for her to leave Peru without any *hasta la vista's*. You know, Simon, I can be quite persuasive when I need to be."

"You are the liar, Manuela. Her sister and brother-in-law spirited her away to get her out of my life. That's what happened. You forget, my dear. My spies tell me everything."

"Yes, they tell me everything too."

Bolivar made a move toward her. "Manuela, my love, your jealousy is ugly. It does not become you. You know I have other women from time to time."

"This was different. I found the poems!"

"So? Poems from a young girl to an old soldier."

"She left, didn't she?"

He leaned forward and said defiantly, "Yes, and I'm leaving, too. *Without you!*"

But Simon Bolivar had already stayed too long in Peru. In Venezuela, earlier that year, General Jose Antonio Paez had called out the militia to put down the roving bands of brigands who were terrorizing the countryside. However, the civil governor of Venezuela refused to acknowledge the order and complained to Santander in Bogota that Paez had exceeded his authority by drafting militia. Thereupon, the Colombian Congress accused Paez of violating the constitution and ordered him to come to Bogota to stand trial. Paez refused. Instead, he expelled the civil governor and took over his functions. The breakup of Great Colombia had begun.

Chapter 48

There was a light drizzle in Bogota when Simon Bolivar rode into the city. It was November 14, 1826. The day was cool and grey. The flags and streamers that festooned the city hung limply. There were no cheering crowds to receive the returning conqueror; only small clusters of people on street corners. They waved half-heartedly. Only occasionally were there cries of "Long live the Liberator!" Bolivar frowned, clearly displeased with the reception.

Close behind Bolivar rode his nineteen-year-old aide and secretary, Belford Wilson, the son of the famous English general and statesman, Sir Robert Wilson. The elder Wilson was a great admirer of Simon Bolivar, as were many liberal Europeans of the period, and had sent his son, Belford Wilson, to South America to serve with the Liberator. Bolivar had made young Belford his aide-de-camp when he arrived in Lima in November of 1823 as a stocky, pink-cheeked boy, a product of Westminster and Sandhurst. Young Belford quickly took great interest in the politics of South America, the wars of independence, and the history of its countries and leaders. He was alert and intelligent and soon became indispensable to the Liberator. He fought gallantly beside him at Junin and was promoted to the rank of colonel on the battlefield.

As if reading his commander's mind, Wilson spurred his horse through the drizzle toward Bolivar. "Remember, sir, you've been away over five years."

Bolivar turned to Wilson. "I used those five years to win tremendous victories. I freed a continent. I defeated the mighty

Spanish Empire. And I am still the President of Great Colombia. I deserve a warmer welcome."

On every street, Bolivar did not fail to notice billboard after billboard, with the same message: "Long Live the Constitution!" He turned again to Belford Wilson. "Santander has done his work well."

"Maybe you shouldn't have left your army in Peru. Maybe you should have ridden into Bogota with the victorious army of Ayacucho at your back and your famous generals by your side, sir."

"No. This will be settled between Santander and me. Alone."

Inside the Palace of San Carlos, General Francisco de Paula Santander and Simon Bolivar bowed farewell to the few gentlemen who made up the official welcoming committee and entered the executive offices of the palace. Santander's smile could have been painted on. Bolivar appeared amiable enough, but his demeanor had the brittle, diplomatic quality of a man hiding his real emotions. Santander walked into the offices with accustomed familiarity. Bolivar stopped and closed the doors to the hallway, where several dozen officers and prominent citizens still stood.

"Thank you for your lovely welcoming speech," he said to Santander.

"Not at all. I'm flattered you liked it, your excellency."

"Yes, it was the falsest, most hypocritical bunch of garbage I've ever heard in my life."

Santander stepped back. The smile left his lips. "You . . . you offend me, sir."

Bolivar's visage was dark. Santander knew the look, and was noticeably frightened.

"I received your letters," Bolivar began. "They insult me. You actually suggest that I should not trouble myself with governmental problems when I return to Bogota. You further suggest I should, instead, lead an army into Venezuela to restore peace there. What do you think I am? A fool? That would have made me nothing but a general under your command."

Santander backed away again. He opened his mouth, but before he could say anything, Bolivar pounded his fist on the large table in the center of the room. "Your government should have crushed the uprising immediately. The minute you heard Venezuela had made Paez its civil and military chief, you should have acted."

Santander found his voice once more. "How? We had no money. We had no arms. And where would we find an officer willing to lead his men against Paez? Remember, he's a fearless, ruthless fighter. I don't think he's ever been beaten. And you ask me why I didn't crush him? Sir, you're unreasonable. Only one man can challenge Paez—and that's you."

"How about these little revolts all over the country, then? You've done nothing about them either."

"The constitution is clear." Santander felt he was on more solid ground now. "We must act only in accordance—"

"I'm changing the damned constitution. It's too weak. It has no teeth. I'm calling a new Constitutional Convention to consider my Bolivian Constitution for all Colombia."

"I've read your Bolivian Constitution, sir. O'Leary brought it to me on your orders. It won't be popular. Congress won't accept it."

"We'll see." Bolivar was testy. "Now tell me why there's no money in the Treasury. Tell me why this country has been governed so badly while I was away. No! I'll tell you. It's because you mismanaged the country and squandered the foreign loans I made possible by my victories. It's because you're a greedy pig who looted the country blind. It's because—"

Santander stormed across the room, his boots pounding on the wooden floor. "I don't have to take this from you! Who kept you supplied with arms and men during all your adventures in the south? Who—"

"You! Francisco de Paula Santander, the same man who revoked the Enabling Act and deprived me of command of my army. For that, alone, I should never forgive you."

"The Congress revoked the act. I had nothing to do with it." But Santander's words lacked sincerity.

"You control the Congress. While I've been away, you and your liberals have made the constitution your God and the Congress your tool. Santander and the constitution have replaced Bolivar and the liberating army."

The two men glared at each other. Bolivar took a letter from his tunic. He shifted his eyes away from Santander and said, "This is from Antonio Paez. I'll read it to you: 'We must confess that Morillo spoke the truth to you in Santa Ana when he said he had rendered invaluable service to the republic by killing off all the lawyers. But we must accuse ourselves of having left Morillo's work uncompleted.'"

"How dare you!" Santander roared.

"I dare." Bolivar's voice was ice cold.

The two men faced each other. Neither flinched. Santander's expression was one of pure fury—and then fear. Bolivar's reflected unbending resolve. Santander hesitated, then walked away and sat down. He slouched in the chair. At last, his face betrayed his utter defeat.

Now that he had thoroughly crushed Santander and established his supremacy over him, Bolivar knew it was time to heal their wounds for the good of the country.

The Liberator's expression turned to one of satisfaction. "But, come," he said good-naturedly, "we must work together to bind the old wounds and unite the country. You are against the Federation of the Andes. You want no part of a union with Peru and Bolivia. Fine. We'll forget it. You want me to go to Venezuela and confront Paez? Fine. I shall go. I want to change the constitution for another that's stronger. You don't. All right, we shall discuss it. I can be convinced."

Santander looked up slowly. "My friends and I think you want to become a dictator. We think you intend to discard the constitution and rule . . ."

"Dictator? I thought you knew me better than that. The word alone is hateful to me." Bolivar banged his fist on the table once more. "No! We shall rule by law. We shall obey the constitution."

Santander looked thoughtful. "You would agree to a free election of the delegates to attend your Constitutional Convention?"

"Of course."

"You will deal with Paez in Venezuela?"

"Certainly."

"You won't press for a union with Peru?" Santander's voice was apprehensive.

Bolivar shook his head. "No. Not unless the Congress of Colombia wants it."

Santander extended his hand. "I misjudged you. Please forgive me."

The two men shook hands. "We shall rebuild the country." Bolivar's tone was now conciliatory. "Please tell your liberal friends they have nothing to fear from me."

Santander nodded. His face was a mass of confused emotions. He looked as if he wanted desperately to believe the great Liberator—but wasn't quite sure if he could.

Chapter 49

For Simon Bolivar, living in Lima had been pleasant. The villa at La Magdalena was beautiful, servants plentiful, the food excellent, and the ladies of Lima were charming and accommodating. They tumbled over themselves to climb into his bed whenever Manuela was away for a few days. And the speeches by the politicians had reflected the truth: He was the only person who could hold Peru together. If he went north, the country would revert to anarchy. So, he stayed longer than he should have.

He was so preoccupied with maintaining order and stability that he did not sense the undercurrent of dissension among the officials and the people. Colombian soldiers, the same troops which had liberated Peru, were resented. The local people were growing weary of supporting a "foreign" army, an army they now considered to be almost an army of occupation.

For their part, the Colombian soldiers were homesick. They'd won the War of Independence. They'd defeated the Spanish Royalists in Peru. Now they wanted to go home. Several of the officers wrote petitions to their congressmen in Bogota. When some of these reached the desk of General Francisco de Paula Santander, they were heeded, and plans began to take shape.

In Lima, several months later, it was not yet dawn when Manuela Sáenz carefully lit her lamp, then crawled back into bed and read Bolivar's letter for the twentieth time.

My dearest Manuela,
  Your letter delighted me. Everything in you is love. I, too, miss you terribly and long to hold you in my arms. You have made my life bearable. Amidst so much misery, you are like a shining star. Stay with me, dear one.

She smiled and sighed. "Oh, Simon, Simon. How I miss you!"

It had been a long time since the Jeanette Hart incident and Bolivar had been as attentive and loving as his long absences allowed.

There was a loud pounding on the door. Manuela angrily shouted, "Go away!"

The pounding continued. Manuela threw off her covers and walked to the door. "Who is it?"

"Maria Perez."

Manuela opened the door. Her friend, Maria, slid quickly inside. Maria was a pretty, generously endowed matron with a large dimple in her chin. "Do you know what has happened?" she asked breathlessly.

Manuela shook her head.

Still panting, Maria Perez's words came rapidly. "At midnight last night, a Colombian officer, a Colonel Bustamante, had the Colombian soldiers arrest every officer loyal to the Liberator."

"What!"

Maria Perez nodded. "It's a full-scale mutiny. Bustamante's

put all the generals in jail. He controls the troops. It happened suddenly, without any warning."

"General Jacinto Lara?"

"Prisoner."

"General Arthur Sandes?"

"Prisoner."

"He can't be. He's an Englishman."

"He's a Colombian general."

Manuela ran her fingers through her hair. "I know Bustamante. Jose Bustamante. He's Santander's man. Of course. It's a *coup d'etat*. What are his plans now?"

Maria Perez looked at the floor and didn't answer.

"Oh, come on, Maria. I know you've been living with that young Colombian hussar. Tell me."

"Well," began Maria shyly, "Bustamante has promised the soldiers all their back pay and that they'll all go home to Colombia."

"And leave Bolivar's government in Peru without support. What else did Bustamante tell the soldiers?"

Maria was thinking hard. "He told them that the only man who really supported their liberties was General Santander, who, I understand, is the vice president of Colombia."

"He's a snake. Please go on."

"Colonel Bustamante said that the new constitution that Bolivar wants to impose on them will mean a return to despotism. He told them that their only hope was to arrest their officers and go home. He'd be made a general, and Santander would protect them all and . . ."

"Oh, be quiet, Maria. I've heard enough."

A month later, on the brig *Bluecher*, Manuela Sáenz stood by the rail and watched the sea birds swoop down and skim the waves then soar again into the blue sky. Out of the corner of her eye, she saw young General Cordoba walking on the deck. She pulled her shawl tighter around her shoulders and gazed at the horizon. As he came abreast of her, Cordoba said, "It's all your fault. Damn you!"

Manuela turned a furious face to him. "You are a miserable, insipid little man. Why don't you try showing an ounce of courage and backbone. Instead you must blame a woman for your problems."

"I would not have these problems if it were not for you! How do you think I like being shoved on board this damned ship and sent off like a criminal in the night? How do you think General Lara likes it? Or General Sandes?"

"How do you think I like it?" Manuela shot back. "Why didn't you take a stand against the traitors? I'll tell you why. Because you're disloyal to the Liberator. That's why."

"I didn't take a stand because Bustamante's men grabbed me in my sleep and put me in chains." Cordoba was shouting now. "But I think they'd have let me loose if you hadn't made such a damned spectacle of yourself. Writing letters. Haranguing the soldiers. Threatening everybody in sight. Ranting against Bustamante. No wonder he's shipping us all back to Colombia like this. It's because of you. Even the Peruvian Foreign Secretary said you were 'an insult to the public honor and morals.'"

"At least I did something. I tried to get the soldiers to come to their senses. I tried to get the Peruvians to act. And you. The great hero of Ayacucho and Pichincha. What did you do? Nothing. Nothing!"

"You're impertinent and insulting. If you were a man, I'd put a pistol ball right between your eyes."

"Go ahead. You're cowardly enough. And disloyal enough. Shooting women should give you great pleasure. And there's no risk. I'm not armed. But to fight for your Liberator against armed traitors? Not you, Cordoba. You caved in as meekly as a lamb. You . . ." But Cordoba had already turned his back on Manuela and was striding down the deck.

Surrounded by his principal officers and advisers, General Antonio Paez sat on the front porch of a large house in the city of Valencia. The house sat back from the street, with a tree-lined driveway to the front door. Horses came and went. Soldiers in uniform guarded the entrance. Paez had gained weight, and his once sharp features had coarsened. But his eyes retained the sparkle of native shrewdness that had made him such a successful military leader. His lieutenants were swarthy, tough plainsmen like himself, but Paez was doing the talking. "The country will accept me as president of Venezuela as quick as you can say, 'To hell with Santander.' There's not one damned thing Bogota can do about it."

"Santander is having fits," commented a young captain.

"The damned lawyer. To hell with him. And to hell with Great Colombia, too." Paez was obviously bitter. "Have you seen the letters I've gotten from Simon Bolivar? He calls me the liberator of Venezuela, the first citizen of the plains, the hero of the Revolution. Ha! Then he has the gall to ask me to submit to the authority of Bogota."

The men around him remained silent. Puerto Cabello, on the coast, was still in the hands of several of Bolivar's generals and Paez had been unable to take it. It was a thorn in his side.

"He's returned from Peru," continued Paez. A frown crossed his scarred forehead. He was lost in thought. Could Bolivar still make trouble for him? No, he'll have his hands full with those damned lawyers in Bogota. Even if he's heard about Puerto Cabello holding out for him, he won't have time to do anything about Venezuela. The frown disappeared, and the sparkle returned to his eyes.

"He's a sick, old man." The speaker was a young colonel. "Besides, Uncle Antonio, even if he tries anything, you can lick him. You can whip anybody."

"Yeah, chief, whenever you reach for your sword, everybody runs like hell."

"I've heard old Bolivar's even forgiven that bastard Santander and he's doing everything but kissing his feet," ventured an older man in the group.

Paez said slowly, "I've replied to Bolivar. I told him we're staying independent and that we're about to hold our own National Assembly. I made it clear that I intend to serve as president, whether he likes it or not."

"I'm afraid that's the way you have to talk to him, chief," said a captain.

An officer of the guard arrived and saluted. "There's a colonel here under a flag of truce. He says he comes from President Bolivar, sir."

"All right. Bring him here."

A few moments later, accompanied by a group of Venezuelan soldiers, Colonel Belford Wilson approached General Paez and his officers. He wore the uniform of the Great Colombian Army. He saluted Paez.

Paez sat straighter. There was something about this young colonel that gave him pause. He seemed resolute and not inclined to take any back talk. He was clearly representing Bolivar. Paez's eyes betrayed nervousness. Several of his men rose and appeared to be about to menace the Colombian colonel. "Sit down," growled Paez. They sat. Paez now rose and extended his hand to Wilson.

Wilson shook it, then handed Paez a sealed envelope.

"You read it." Paez was getting better and better at reading, but he wasn't fluent. Long words were still hard for him, and Bolivar always used a lot of them. Besides, he didn't want his men to suspect he wasn't completely literate yet.

Wilson opened the letter. "It's from the president of Great Colombia. It starts: 'In your correspondence, you have been calling me 'Citizen Bolivar.' I wish to make things clear. I am still your president and will be addressed accordingly.'"

Paez simply nodded. His face reflected no emotion. He was waiting to hear what the Liberator's intentions were.

"He's heard you are holding a national assembly, and he tells you anything it does will be declared null and void." Wilson's harsh tone of voice caused those present to remain silent and somber. They were listening intently.

"I will read the Liberator's words again: *'You seem to feel I am a simple citizen without any legal authority. I will set you straight: There is no lawful authority in Venezuela but mine, and my authority is supreme. If you do not believe me now, you will when I have you hanged for high treason.'* " Here Wilson stopped and looked around the porch to let the words sink in before he continued. "'I have come to save you from the crime of civil war. I have come so that Venezuela may not again be stained with precious blood. At the same time I must remind you: The will of Providence sends my enemies to ruin, but good fortune shines on my friends. So, my dear Paez, it is up to you to decide now on which side you care to place your destiny.'"

Wilson folded the letter. Silence greeted the final words.

Paez bit his lip. "Where is the Liberator now?"

"He's on his way here." Wilson spoke coolly. "With the Great Colombian Army at his back. He's already cut you off from the plains. So you're caught like a rat in a trap between Bolivar's forces in Puerto Cabello and the Liberator, himself, marching on you with his army."

"No problem for us . . ." began one of Paez's officers.

"Shut up!" snapped Paez.

He looked directly at Wilson. "Listen, Colonel, I begged the Liberator to mediate between Santander and me. All I asked was for him to be fair. But, instead, he's agreeing with that damned Santander on everything. He's siding with that bloody crooked lawyer. Why? Why the hell is he marching against me?"

"Because you have seceded from Great Colombia. That's why he's marching against you."

Paez pulled out a handkerchief and mopped his forehead. "If the great Bolivar is on his way here with an army, he means to fight us, all right."

"Don't worry, sir, he's a sick old man," said a swarthy young colonel.

"Shut up!" barked Paez again. "Sick old man, my ass. Morillo thought he'd beaten him time after time. La Torre underestimated him and got licked at Carabobo. I was there! Pasto fell to him. La Serna and Canterac never thought he'd even be able to get into the Andes. And what happened? He beat the hell out of every single one of them. That's what."

"He gave in to Santander," ventured another officer.

Wilson cleared his throat. "The Liberator has placated Santander because he doesn't want to leave an enemy at his back while he's crushing you, sir. He will take care of Santander when he returns to Bogota."

The plainsman nodded. He walked a couple of steps over to Wilson's side and patted him on the back, then took his arm. "You know, Colonel, I joined General Bolivar of my own free will way back in '17. The Liberator and I were comrades in arms for a long time. We fought against the Spanish together like brothers. We shared some pretty rough years. I think it's time for two old friends to sit down and straighten out their recent misunderstandings, don't you?"

Chapter 50

Wilson rode beside Bolivar as they approached Caracas. The road was old and rutted, edged on both sides by foothills of green grass. Neither man spoke until Wilson asked, "Do you really think it's wise to meet General Paez alone like this, without an escort?"

"It has psychological value. I met the great Morillo the same way. It impressed him. It made my diplomatic victory at Santa Ana possible. I rode into Pasto alone after Bomboná. I entered Lima with no more than a dozen lancers. A commander in chief riding alone and unprotected into the midst of an enemy army makes an impression that the world will remember."

"Sir, what do you really think of General Paez?"

"I think my friend, General Paez, is the vainest, stupidest, most ambitious man in the world. His pride and ignorance blind him to his limitations, and he will always be a tool in the hands of his advisors. *The man knows only his own nothingness.*" Bolivar knew he was exaggerating.

But his feelings were clouded. Paez had been his ally when he didn't have many. He had supplied horsemen to cross the Andes and liberate Colombia. He had fought like a madman at Carabobo, lost his best friend in the battle. But now . . .

"You're exceptionally angry at him, aren't you?"

"Wouldn't you be? I'm not a well man, Belford. Yet I'm back in a saddle, back in uniform again. I had to raise an army just to come here to straighten out Uncle Antonio Paez, when I should be attending to the mess in Bogota. I only hope Santander will behave while I'm gone. He has his own political party now and keeps the people stirred up about their precious 'liberty.' He insinuates I'm going to take it away from them. Hell! I gave it to them in the first place. Have they forgotten that? Santander worries me to death."

"I guess that's why you wanted to settle things so quickly here. But, sir, do you think that was wise?"

"Paez accepted my authority as president and Liberator and promised to obey all my future orders. What more could I ask?"

"Sir, he would have accepted any terms you dictated. He certainly didn't expect you to be so generous."

"Belford, Venezuela is back in the fold of Great Colombia. Paez has abandoned his plans for an independent national assembly. I have averted a full scale civil war. I've achieved peace and unity. What more could I want?"

The church bells of Caracas were still ringing; the people still cheering; and the flags still flying in the breeze, when the great Liberator ascended the steps of the Cathedral with Antonio Paez at his side. Both men were smiling, Paez shyly, Bolivar joyously, obviously delighted with the reception. Looking out on the arch of triumph, the garlands of flowers, the flags and the palms, he raised his hands and beamed at the crowds. Then, overwhelmed by the occasion, he motioned for the crowd to quiet down so he could address them. A gesture was clearly called for.

When the people had become silent, the Great Bolivar pointed to Paez, standing at his side. "People of Caracas and of all Venezuela: I give you the savior of the republic, the man who has preserved the ship of state from catastrophe."

As the crowds cheered, General Laurencio Silva, one of the few officers who had ridden into Caracas to aid the Liberator should he need it, turned to Belford Wilson. "Damn! He's going too far. I think he should have hanged Antonio Paez for treason, and here he is, praising him to the skies."

Wilson nodded. His face was solemn. "All hell's going to break loose when they hear about this in Bogota."

The Liberator wasn't finished yet. After silencing the crowd, he put on his most ceremonious visage and, slowly unbuckling his sword and scabbard from his belt, he announced, "This is my sword, which has seen me through two hundred bloody battles, the sword of victory. But now we are at peace, and I honor the hero of Venezuela by presenting him with this token of my great esteem." With this, the Liberator solemnly handed his sword to Antonio Paez.

After the cheers subsided, General Paez raised the sword above his head with both hands. "My fellow countrymen, Bolivar's sword is in my hands. For him and for you, I shall march with it to Eternity!"

In a corner of the palace's large formal reception room in Bogota, Francisco de Paula Santander slouched in his chair. He held several handwritten pages. Dr. Vicente Azuero, a malevolent ferret-faced man who had been a fierce opponent of Bolivar ever since that day in 1819 when the Liberator snubbed him during his triumphal entry into Bogota, walked up and down in front of Santander. Joaquin Mosquera, the former Colombian ambassador to Peru and still a friend of both Santander and Bolivar, sat beside him looking at the floor.

"I simply cannot believe this," said Santander. "The Liberator has actually given his approval to Paez's rebellion. He has sanctioned treason."

Mosquera looked reflective. "I don't think it's that bad. After all, Paez submitted."

Santander looked at Mosquera as if he were a backward child. "Paez has submitted. That is correct. But submitted to what? The constitution? No. To the civil authority of Bogota? No. To the will of the people? No, not to any of these. Antonio Paez has submitted to Simon Bolivar, personally. And that's all."

Chapter 51

Sitting on his favorite couch in the Administrative Hall of Caracas, Simon Bolivar looked extremely ill and unusually sad. "I just can't believe it," he said to Daniel O'Leary who was sitting behind a desk sorting through mail. "How could Bustamante do such a thing? He's arrested all of my Venezuelan and English officers and has already left port with his troops for Guayaquil."

"And we've received some more bad news, sir. As soon as Bustamante left, the Peruvians abrogated your constitution and elected a new president."

Bolivar shifted his position slowly so he could look directly at O'Leary. He couldn't comprehend the fact the Peruvians had behaved so badly.

"The Peruvians didn't like your leaving them, sir. They didn't understand that you have problems in Colombia. And they resented the fact you left so many Colombian troops in Peru. It was a burden for them. A foreign army on their soil."

"It will end in anarchy," said Bolivar. "If I didn't have to contend with Santander, I'd go south again and punish this little insect. But if I go south, the north will revolt; if I stay in the north, the south will disintegrate."

"Did you hear what happened when the news of Bustamante's revolt reached Bogota? Santander had all the bells of the city rung. He went out into the streets, and the crowds cheered."

"Stop. These events convey only one thing: There's been a violent reaction against me, and not only against me but against my political concepts. This doesn't anger me, Danny. It grieves me. I'm going to resign the presidency of Great Colombia."

O'Leary spoke carefully. "Every time you've done that, sir, the Congress and the people have begged you to stay on. And you think that will happen again. But this is different. You have a powerful adversary now. Santander is a leader. And he is openly opposing you. He and his friends have founded a newspaper and they're demanding your removal from the presidency."

As O'Leary watched with pleasure and wonder, the sparkle returned to Simon Bolivar's eyes. Ever since the defeat of the Spanish royalists, he'd had no worthy enemy. Now he did. Santander! He pounded his fist into his open palm. "No, dammit! We'll fight. I once saved my country by declaring *War to the Death*. And to save it again, I'll fight again, even if I die by the traitors' daggers."

In the great room of the palace at Bogota, Dr. Vicente Azuero was agitated. He leaned forward and rested both arms on Santander's desk. "He's on the way back," he told the general, his words tumbling out in fear. "Do something."

"I've written to him," Santander responded. "I've told him that his coming here with his army is a breach of the constitution. It's like Napoleon returning from Egypt." But Santander knew his letters would have no effect at all on Bolivar.

Azuero slapped his hand on Santander's desk. "He'll show his true colors now. He'll take over by force of arms. Even though these dolts have elected him president again, he will never take the oath of office on the constitution. He'll seize the government and declare a dictatorship. That's what he'll do."

Azuero was wrong. Simon Bolivar returned and took his oath of office swearing to obey the constitution and pledging himself to rule democratically as long as he held the office of president.

At General Francisco de Paula Santander's house, dinner was over. The ladies had left for the drawing room, and Vicente Azuero, Joaquin Mosquera and Santander sat at the table drinking port while the servants cleared away the plates.

Azuero spoke first. "Well, what now? We tried every trick in the book to keep him and his army out of Bogota. Nothing worked."

"The Liberator's prestige is too great," said Mosquera. "It's overwhelming."

Vicente Azuero smiled, showing a row of straight, white teeth. "Therefore, my dear Joaquin, we must use every means at our disposal to destroy that prestige. *We must drag him down to the level of mortal men.*"

In Quito, at the home of her half-brother, Jose Maria, Manuela Sáenz rose from her hammock. She had heard the clatter of a horse's hooves and wanted to see what was happening. She immediately recognized the rider. "Arthur!" she cried and clapped her hands together.

General Arthur Sandes, tall, blond and well-scrubbed, dismounted, strode over to Manuela and embraced her. "Good to see you, Manuela. We haven't seen each other since that beastly ship, have we? Here, I've come to drop off this letter."

"Who is it from?"

"Who do you think?"

Manuela grabbed the envelope and tore it open. She looked up at Sandes. "Excuse me. I haven't heard from him since I left Lima."

She walked a few steps and read the letter.

> Manuela,
> The memory of your enchantments dissolves the frost of my years. Your love revives a life that is expiring. I see you always, even though I'm far away from you. Come. Come to me. Come now.

She read the letter once more, then held it to her bosom with a smile on her lips. When she returned to General Sandes, she was beaming. "Arthur, he wants me. He wants me to come to Bogota."

"I know. I've brought a squadron of lancers to escort you."

"That's not necessary. I can always get around by myself."

"No." Sandes was firm. Then, he smiled. "Things are unsettled.

It's not safe. Lots of robbers and brigands and that sort of thing, don't you know."

"Will you take a letter to the Liberator for me?"

"Rather. He'd have me shot if I didn't."

Manuela wrote rapidly:

Dearest,

I am very angry and very ill. How true the saying is that long absences kill little loves and increase great passions. You had a little love for me, and the long separation killed it. But I, I who had a great passion for you, kept it to preserve my peace and happiness. And this love endures and will endure as long as Manuela lives.

I am leaving for Bogota as soon as I can assemble my things—and I come because you call me to you. However, once I am there, do not afterwards suggest that I return to Quito.

She sealed the envelope and handed it to Arthur Sandes. "Now, Arthur, let's have lunch."

On the outskirts of Bogota, in a hollow between two soaring mountain peaks, sat a small jewel of a house called *La Quinta*. It was a one-story colonial cottage with a red tile roof and white walls. Its surroundings were lovely: stately cedar trees ringed the outer boundaries and a profusion of flowers clustered around the house. It was a gift to the Liberator from the citizens of Bogota, presented to him after he liberated Colombia.

Late in the afternoon Colonel Daniel O'Leary pulled up his horse at the gate in the wall surrounding La Quinta. He surveyed the property, as a sentry amiably asked his identity and purpose. "I'm Colonel O'Leary, a friend of the Liberator."

"The Liberator is expecting you, sir." The sentry gave him a smart salute.

Bolivar had been using O'Leary as a high-level messenger to leaders both within and outside of Great Colombia, and he was glad to be back. Entering the house, O'Leary noted that the furniture was aged mahogany, creating a warm atmosphere. The house consisted of only a few rooms built around a large patio: a library, living room, dining room and bedroom, each with its own fireplace. In the living room, the Liberator stood talking to several other officers. But as soon as he saw his old friend, he detached himself from them and went to greet O'Leary. They embraced, both smiling. "This place is magical," said O'Leary.

Bolivar nodded. "I prefer it to the palace and spend as much time as I can here. But come. I think you know everybody, William Fergusson." Here Bolivar waved towards the sandy-grey haired Scot wearing the uniform of a colonel, "Jose Maria de Cordoba," another wave, "Rafael Urdaneta, Doctor Moore."

They all responded by greeting O'Leary with a warm embrace or a shake of the hand, although the pleasantries lasted only a minute or so. "Let me bring you up to date." Bolivar was speaking to O'Leary, but his words brought silence to the whole room. "I have decided to hold the Great Convention in the town of Ocaña, away from Bogota. There will be less distractions there, less pressure on the delegates."

General Urdaneta spoke up. "The Liberator was just telling us that only the best men available should attend the convention. Men who represent the country."

O'Leary nodded. "We shall make sure that's what happens." His tone was matter of fact. "We shall pick the delegates with care and . . ."

"Wait! That's the problem I was getting to," broke in Urdaneta. "The Liberator has informed us that neither he nor any officials of the government can engage in the campaign. And I say that's ridiculous. You know Santander is going to campaign. And he wants only liberals to attend the convention."

"As far as campaigning goes," said Bolivar, "my reputation won't permit me to take any part in the election proceedings. I don't want to be accused of using the executive power to further my personal interests. That's the reason I've instructed all government officials to refrain from trying to influence the voting."

"Santander has no such scruples." Urdaneta was becoming angry. "He's going to do everything in his power to get as many liberals elected as he can. He's out to control your Great Convention, sir. He wants to destroy you."

"The people will see through him."

"I wouldn't count on that, sir." O'Leary had been riding all over Great Colombia and knew the people's mood better than his leader did. "He has become quite popular, you know. He has convinced a lot of people that you want a new constitution so you can rule as a dictator, and that Santander is the only one who can protect their liberty."

"No," interrupted Bolivar. "The people know that if I wanted to be a dictator, I'd be a dictator. Everybody knows that all I want is a strong constitution."

"We know that, sir," said Cordoba. "But the people don't. I think you should explain what you want to do, so they'll understand it. Otherwise, they have no choice but to believe Santander."

Bolivar threw a haughty glance at his young general. "I cannot stoop to that. I am their Liberator. I can't turn around now and beg them to do what I want."

"You have only two choices." O'Leary picked his words carefully. "Either be a dictator and rule by military force—or use your personal magnetism and your powers of persuasion to win over the people and govern effectively."

"No! Neither," cut in Bolivar. "I refuse to be a dictator under any circumstances—and I also refuse to lower myself to beg. George Washington refused a crown and refused to be a dictator. So do I! And Washington was elected twice, *unanimously* . . . ."

He was interrupted by the arrival of Belford Wilson. Bolivar greeted him warmly. Wilson was smiling as he embraced the Liberator. "I see you have the loyal old guard here. If Marshal Sucre weren't president of Bolivia, he'd be here, too, I imagine."

"I almost wish he were," said Bolivar.

"And Manuela?" asked Wilson. His voice was low and his words tentative, yet a feeling of tension fell over the room.

"She is on her way to Bogota, at my invitation," replied the Liberator.

"No!" Young Cordoba actually gasped. "Not here. I was with her on the same boat as Sandes and Lara. She's trouble. Things are delicate as hell here, anyway. All we need is for Manuela Sáenz to stir things up! Then we'll really have a mess on our hands."

"Oh, go on," said William Fergusson. "She's delightful. She'll be a great addition to."

"I'm not sure." O'Leary was shaking his head. "I happen to like her very much, but she's not always diplomatic."

"I have not yet had the pleasure of meeting the lady," said Urdaneta. "But I understand she is gracious and kind and."

"She's a fury!" declared Cordoba. "She should stay in Quito, put away in a convent."

"Enough!" said Bolivar. "I need her, and she's coming. I'm surprised that gentlemen would even discuss a lady this way. Shame on you. You can disagree with me on any matter you want, including politics, but, in Heaven's name, leave Manuela out of it."

The men fell silent. But their faces expressed different emotions. Some were troubled, some concerned, others wary. Only Fergusson seemed frankly cheerful.

Bolivar refused to explain his bond with Manuela. At times, it was difficult even for him to understand.

Bringing Manuela to Bogota was a terrible risk but he felt he must have her there. His life, it seemed, was like a vacuum without her. There was no one else with whom he felt the same pleasure, the same deep connection. They were, he felt, in perfect step, but always out of step with everyone else.

Chapter 52

Several days after General Arthur Sandes had delivered Manuela's letter to La Quinta, his squadron of lancers was escorting her from the hacienda outside of Quito to Bogota. It was a difficult journey along the eastern foothills of the Andes, taking over a week. They stopped in numerous villages along the way and in each one Manuela encountered anger and disillusionment over Bolivar. At one point, she asked the squadron's commander, "How can the people be so hostile to their Liberator?"

The commander pointed down to the old bridge they were crossing. The wood was rotting; the paint had faded. Stones from the foundation had been taken to build huts. Soon the span would crumble into the river. "When we were a colony of Spain, bridges like this were always maintained in perfect condition."

Similar scenes of decay greeted them at every turn. The people of Great Columbia were poor and disheartened, their lives unraveling with each passing day. On their houses the paint was peeling, the woodwork rotting. Their roads were almost impassable. The market stalls were empty. Parts of public buildings were falling to the ground from lack of repair. For the common people who had supported the revolution, the price of freedom was far too high.

Arriving at La Quinta, Manuela was relieved and happy to see its absolute beauty. The house was newly painted; the grounds were immaculate. She dashed in the front door, desperate to see her Bolivar but was disappointed to find only the servants. "The Liberator has gone to the city," they told her.

It was not until sunset that Simon rode wearily into the courtyard. Looking out the window, Manuela saw that he was alone and shook her head in frustration. He can't conceive of anyone physically attacking him, of trying to kill him, she thought. He's *The Liberator*. He doesn't need an escort. She smoothed her white dress and made sure her bodice was half unbuttoned so that when she breathed deeply, her ample breasts were almost totally exposed. Then, she placed herself in the center of the small living room, holding a red hibiscus in her hand.

When Bolivar caught sight of her, his face lit up as if by magic. He held out his arms and gathered her into them. "I'm so glad you're here," he said softly.

"You talk too much," said Manuela. She was breathing enticingly into his ear.

Bolivar took her hand and they walked slowly towards the bedroom. Manuela stopped and unbuttoned the remaining buttons on the front of her dress, pulled the ribbons which held it up at the shoulders, and let the frock fall to the floor.

Simon looked horrified. "The servants!" he whispered.

Manuela smiled archly. "I gave them the night off."

A few seconds later, they were making passionate love on the large vicuña rug in the middle of La Quinta's living room.

In an office in the palace in Bogota, Belford Wilson placed a newspaper in front of Colonel O'Leary. "Look! How much longer can the Liberator put up with this kind of thing?"

O'Leary read aloud, "A MISTRESS IN THE PALACE AND SOLDIERS IN THE STREETS! Is this the independence we sacrificed so much for?" He frowned and said, "They are living together openly. Here in Bogota they prefer these things be done more discreetly."

"They are hypocrites," Wilson asserted. "And the Great Bolivar isn't."

"Anyway, she's taking good care of him. Have you noticed how much better he seems since she arrived?"

Wilson nodded, then resumed sorting through the newspapers and broadsides. "Look. Here's another article. It says, 'Sinister plans of Bolivar. Reign of terror will make Morillo's seem mild by comparison.' How can they say things like that? They know they're not true."

"The worst of it is, the Liberator won't let us fight back. And, remember, Danny, it's the Liberator's own fault. He's the one who decreed 'Freedom of the Press.' Well, now he's got it—in spades."

O'Leary seemed pensive. "I'm afraid our great Bolivar has become a disillusioned man. That's why he wants the president appointed for life with unlimited powers."

Wilson said slowly, "I wonder sometimes if, maybe, Santander might have a point. He wants a constitution that gives the power to an elected Congress."

O'Leary shrugged.

"And Santander wants the Congress to govern the country. They're too far apart, I'm afraid. I don't see any middle ground."

* * *

The results of the elections of delegates to attend the constitutional convention at Ocaña were in. Standing in his large office, Simon Bolivar, the President of Great Colombia, was enraged. The blood vessels swelled on his forehead; his temples pulsed. "How could they do this to me?" he asked.

O'Leary, Urdaneta and Cordoba stood in front of him. None answered.

"Santander is clearly the idol of the people," Bolivar said with chagrin. Then, his face brightened. "No! My enemies arranged a fraudulent election. They won by cheating."

O'Leary cleared his throat. "Excuse me, sir, but that is not possible. You control the government."

"The pity is, you could have won so easily," said young Cordoba. "If you had only used your power. If you had allowed the mayors and the priests and the members of the government to get out and campaign, you would have won every single delegate."

"I scorn such practices."

"Santander damn well doesn't scorn them," shot back Cordoba. "He used them, and he won."

"It proves he's nothing but a demagogue." Bolivar spoke scornfully. "From now on I won't even mention the constitution in my public addresses. We'll simply hope for the best."

Alone, Simon Bolivar reflected on his complete political defeat. He knew he had been arrogant in thinking that the people would vote for him simply because he was their Liberator, the founder of their free republic. He took it for granted that they would see through the machinations of Santander's demagoguery. He wished that he had fought the political campaign as he fought a military battle—with all his might and heart! But instead he had allowed his enemies to overcome him. Now, he realized he had to salvage what he could. But he was hurt more than he could admit, even to himself. The people had not backed him.

Chapter 53

The Convention of Ocaña was dissolved when Bolivar's people pulled out and took enough other delegates with them to destroy the quorum, thereby effectively nullifying the elections. Bolivar's friends immediately arranged for him to be called upon to take over all the powers of government "in order to save the country." But the damage was done. By invalidating the elections, Bolivar seemed to confirm exactly what his enemies were saying—that he was going to rule as a dictator and ignore the constitution completely. Many of his former staunch supporters began having doubts about the Liberator's motives. In effect, in the eyes of most of the citizens, he had retained power by executing a *coup d'etat* against the constitutional representatives, elected strictly by the rules he, himself, had decreed.

Several weeks later, Dr. Vicente Azuero, the fiery, grey-haired adherent of Santander, was putting up wall posters around the city, proclaiming, "No Liberty under Bolivar!" When he stepped back to admire a freshly hung sign on the wall of a house in downtown Bogota, a large, young soldier came around the corner, saw him, and immediately knocked him to the ground. Suddenly the soldier was all over Azuero, hitting and kicking him senseless. From the opposite direction, General Cordoba came walking down the street. He saw what was happening, broke into a run, and drew his sword. "Stop that!" he shouted, confronting the soldier.

The soldier looked abashed. Azuero slowly got to his feet. Cordoba gave the old man a contemptuous glance. "You! Stop putting up your filthy signs and get out of here! Now!"

Azuero staggered off, visibly shaken. Cordoba put the point of his saber against the soldier's chest. "Who put you up to this?"

The man's eyes widened in terror. "Nobody," he gasped.

"If nobody told you to find Dr. Azuero and beat him up, then I'm going to drive this saber right through your heart—now." As he spoke, Cordoba pushed the point of the weapon firmly into the man's chest, causing him to step backward until he hit the wall of a building behind him. "Now or never," said Cordoba. "Who told you to do this?"

The man swallowed hard. "The lady Manuela."

General Cordoba wasted no time in reporting the incident, personally, to Bolivar. He stood in front of the Liberator in the Palace of San Carlos and spoke forcefully. "You've told me the situation is most delicate. You've told me we must do nothing to incite the people, nothing to disturb the public tranquility. We must deal with the liberals diplomatically, so they'll work peacefully and respect your rule. Am I correct?"

"Of course."

"So, what are you going to do about the incident with Dr. Azuero? That is exactly the kind of thing I thought you were trying to avoid."

"It is. The entire affair is reprehensible."

"The people all hate Manuela. The country is poor to the point of starvation, and she spends money like water. Paris gowns. Expensive perfumes. Wines. Parties."

"Enough!" Bolivar was annoyed. "I shall see that she behaves. But I cannot send her away. She is necessary to me."

Cordoba was about to reply, when Colonel O'Leary entered. "I have another unfortunate incident to report," he said, seemingly reluctant to go on. "You know that scandal sheet they like to call a newspaper? The one Azuero runs?"

Bolivar and Cordoba both nodded. "What's it saying now?"

"Well, it printed an edition calling for 'Death to the Tyrant,' and when William Fergusson saw it, he went right down to the paper's offices and beat up the editor, wrecked the office and smashed their press."

"My God," said Bolivar. "My friends are destroying me."

"The opposition, as they call themselves now, want an apology for these acts of terrorism and persecution."

"Yes, of course."

"They want Colonel Fergusson court martialled and dismissed."

"I'll reprimand him. But I won't dismiss him. We shall handle it ourselves." Bolivar looked pensive. "This kind of thing has to stop. We have won back the government. We have overcome the opposition. We're in charge. We promised not to seek revenge. We declared there would be no recriminations. We must respect those promises. Our policy must be reconciliation. Ignore their damned lying, blasphemous papers and pamphlets. Let them say what they will. Our task is to rebuild the nation, not tear it down."

The United States of America, having recognized the government of Great Colombia, sent its first diplomatic representative to Bogota in 1828 in the person of William Henry Harrison. Harrison had been a general in the Indian Wars and the War of 1812 and was famous as the victor of the Battle of Tippecanoe. He had also served as Secretary of the North-West Territory, a Delegate to the Congress and a U.S. Senator. Being more politician than soldier, Harrison fell in with the Santander faction, who took advantage of his naivete concerning the local situation to prejudice him against Bolivar. At a formal dinner honoring the father of his country, Harrison toasted George Washington with the words, "One George Washington dead is worth a hundred Simon Bolivars alive."

An account of the affair, naturally, got back to Bolivar's people. "You love the United States so much. What do you think of them now?" asked Cordoba.

"Yes," said O'Leary. "What about it? The man has insulted you, sir. You must ask for his recall at once!"

Bolivar shook his head slowly. "No. I don't want to ask the United States to withdraw the first ambassador they've sent us. He doesn't represent the feelings of his countrymen, I'm sure. Only his own. If I hadn't been so busy, I'd have spent more time with him and he'd have been *our* man. But I didn't, so we must accept what has happened."

It was not necessary for Bolivar to do anything. President Andrew Jackson removed Harrison almost immediately. Harrison had been in Bogota less than a year.

Chapter 54

It was early in the afternoon, and the outside of La Quinta was festooned with flags and patriotic banners. Tables were placed on the lawn. A party was in progress and the guests included many officers of the army and many civilian members of the council and cabinet. Manuela Sáenz, basking in her role as hostess, greeted them all with flirtatious charm. "Where is the Liberator?" asked an arriving guest.

"Oh, he's working. Arranging a foreign loan or something. You know how he is."

"But it's his birthday."

"That's why we're celebrating. He's forty-five today." Manuela laughed, clearly in high spirits. Then, she caught sight of Colonel Fergusson making his way from the punch bowl. Detaching herself from the others, she went over to him. "I hear you've been a naughty boy."

He bowed sheepishly. "Lost my bloody temper. Those damned scoundrels. I don't see why the Liberator lets them print all that garbage."

"He gave you hell, too, eh?"

"He could have done a lot worse."

Manuela laughed again. "Santander demanded that you be discharged. But, do you know what? That made the Liberator so mad, he's decided to make you a general!"

"Jolly good." Then, Fergusson leaned over and whispered confidentially, "Make the Liberator take care, my dear. Now they'll try to kill him, you know. It's the only way they can defeat him."

Manuela sighed. "I know. But he simply won't believe me."

The guests were thinning out. Among those who remained, the talk and laughter were loud. Manuela was among the revelers, an obvious favorite of theirs. She had always been fond of her liquor, and in the past could match the strongest lancer drink for drink. Lately, though, when she had had too much port, her voice became loud and her actions erratic, even for Manuela.

Just outside the walls, the Grenadier Battalion, which had entertained the guests earlier with their precision drills, could be heard joking with each other while they drank their rum and ate their black bread.

As usual, the foreign officers clustered around Manuela. They were all talking in loud voices, obviously feeling the effects of the wine that was still being served. William Fergusson was unsteady on his feet. "Hey, Manuela, is that bastard, Santander, still vice president, or what?"

"I heard the Liberator abolished the office of vice president," said a Colonel named Crofton.

"Sure," ventured a young officer, who wouldn't have dared speak to his superiors if he hadn't been drinking all that wine. "But the liberals claim Santander is still the legal vice president, anyway. If I were the Liberator, I'd shoot the bastard and be done with it."

"That's just what we shall do," shouted Manuela. "Jonathas!" she called to a servant. "Get me some sacks."

In an instant, two cloth bags were stuffed with old rags until, placed one on top of the other, they were roughly the height of a man. An officer's old, frayed tunic was hung on the effigy. A string was tied near the top to form a head and neck. A mustached face was painted on it, and a sign placed over the dummy's chest that read: "FRANCISCO DE PAULA SANTANDER—EXECUTED FOR TREASON." By the time the mannequin had been propped up against a wall, Colonel Crofton appeared with a firing squad of

Grenadiers. One of Manuela's servants had dressed in a black cape and, pretending to be a priest, gave the dummy the last rites of the church. Crofton told the lieutenant in command of the firing party to prepare to fire. But the man, who seemed to realize for the first time what was going on, suddenly put his hand to his mouth in dismay. "Sir! I can't take part in anything like this. It's a farce."

"You're dismissed," said Crofton. "I'll do it myself."

Then, Crofton raised his sword and commanded the soldiers, who still stood at attention. "Ready!" They brought their rifles up. "Aim!" Then, swishing down his sword, Crofton yelled, "Fire!"

The volley rang out; the dummy flew apart under the impact of the rifle balls.

"It was totally senseless. One afternoon's indiscretion, and all my plans are ruined!"

Bolivar was as angry as Manuela had ever seen him. He darted around the living room of La Quinta throwing pillows and books, his true temper unleashed. Manuela stood in the corner, recoiling beneath his wrath. Suddenly, he turned to her, seemingly ready to pounce, and shouted, "How could you possibly have been so stupid?"

Manuela glared back, her own spark ignited. "How dare you speak to me that way! It wasn't a public crime. It was a private party."

"It was insanity!" Bolivar boomed.

"Who told you about it?" Manuela demanded, obviously trying to change the direction of the conversation.

"Everybody in the city knows about it. The shot was heard all over Bogota."

"Who told you?" persisted Manuela.

"Cordoba."

"He would. He hates me."

Bolivar fell onto the couch, coughing, clearly exhausted by his own furor. "I apologize for your stupidity," he said softly. "I told him I'd do something about Crofton."

Manuela inched toward him, sensing an open door. "And what do you intend to do about me?"

"My friends insist I banish you." Bolivar's words were tired, without emotion.

Manuela breathed deeply. "And?"

Bolivar shook his head, his anger completely dissipated. "I cannot banish you from Bogota because I promised you I wouldn't. I have tried to break with you, Manuela, but I can't." She moved closer but he held up his hand. "You are positively hazardous but for reasons I don't understand, I can't seem to live without you. We both know that."

Manuela looked on silently.

"But I'm going to have to send you away from La Quinta for a while. This last escapade is far too serious to overlook. It's going to cause problems."

Manuela tossed her head. "Suppose I won't go?"

"You'll go."

She looked into Bolivar's eyes for a glimmer of weakness—or compassion. She saw only resolve. She bowed. "Yes. Because you are my master, I shall obey you."

Chapter 55

General Cordoba and Colonel Fergusson, together with Urdaneta and O'Leary were gathered around Bolivar in the palace office. "Look at these." Fergusson displayed the pamphlets which proclaimed "Death to the Tyrant!" and "There is no liberty as long as Bolivar lives!"

"We know there are plots to kill you," Urdaneta said quietly, keeping his emotions under control.

"Give us the authority, sir," said Cordoba. "We'll break this thing wide open and bring the damned assassins to book."

"No. I want less dictatorial decrees, not more."

O'Leary spoke sternly. "If you don't take some sort of precautions, we shall have to take them for you."

"What do you want me to do?"

"Discontinue your daily horseback rides, unless you take an escort. Continue to live here in the palace. Double the guard."

Bolivar waved his hand in disgust. "What do you think I am? A coward?"

"No, sir," said O'Leary. "But we think you're a reasonable man who will be guided by common sense. This is an emergency, sir!"

"And General Santander?" asked Urdaneta.

"I have named him ambassador to the United States." Bolivar thought of this as an honorable banishment. "He'll be leaving the country."

"He hasn't left yet." Cordoba reminded him. "And you can't tell me the plot doesn't center around that gentleman."

The Liberator sat alone in the large room in the palace with one leg over the arm of the chair. In his hand was a note from Manuela. He was still upset about the way his subordinates had talked to him, but he realized it was their concern for his safety that made them address him in that manner. He smiled as he re-read the note from Manuela.

> My dearest,
> Even now, you refuse to believe there are plots, but there are. And I shall protect you with my life, if I have to. I love you. But I shall not come to your house again until you ask me to come. I hope you want to see me soon, though.
> As ever, your Manuelita.

Bolivar ran a hand across his forehead. He sighed.

Several days later, Generals Urdaneta and Cordoba, joined by Colonel O'Leary, stood grim-faced before the Liberator, who sat back in his chair staring at the reports on his desk. Cordoba spoke first. "An officer named Triana got drunk and started raising hell in the barracks. He was yelling that 'Bolivar's tyranny would be drowned in blood.' He claimed that he and his friends were going to do in 'that old man Bolivar.'"

"Are you convinced now, sir?" asked Urdaneta.

Bolivar nodded. "This man, Triana, was drunk. But the situation is plainly serious. The plots appear to be real, I must admit. And, gentlemen, I am no saint. I have no desire to be a martyr. We must use caution. We have to take steps to protect ourselves."

The men standing before him sighed with relief. "Thank God," breathed O'Leary.

"Shall we round up all the people we believe are plotting against you, sir?" asked Cordoba. "Horment, the Frenchman? I think he's a Spanish spy, anyway. Azuero?"

"Santander!" exclaimed Urdaneta, voicing the name they had all been avoiding.

"No." Bolivar pounded his desk. "They haven't done anything. There is no legal basis for taking them into custody."

"What do you want, sir?" Cordoba's eyes blazed. "Do you want to wait until they've killed you?"

"They won't kill me. I don't think they have the guts. Anyway, we're prepared, or at least, we shall be in the next few days."

"Yes, sir," came the response in unison.

After his friends had left, Simon Bolivar took a blank sheet of paper and wrote on it, "Come. Come to me. Come now!" He signed it and dated it "September 25, 1828."

If there were plots, Bolivar suspected the traitors would kill Manuela, hoping that they would get away with it because so many people hated her. But Bolivar was determined to protect her.

Manuela Sáenz stood beside a table in the middle of her second floor living room in Bogota. She was carelessly dressed but, as always, she exuded sensuality. Her large brown eyes and full lips stood out against her pale skin, and her shiny black hair seemed to sparkle in the candle light. At thirty, she was in her prime. In her hand she held a note. She read it again. "Come. Come to me. Come now!" It was signed, simply, "Bolivar."

"He's refused to see me for over two months. Ever since I had that silly Santander effigy shot. Don't you think I have any pride?"

Jose Palacios was waiting expectantly. He was dressed in civilian clothes but his outer coat consisted of three Spanish Army overcoats sewn together. He seemed agitated. "You must come. He needs you."

Manuela nodded. "If he sent you, it must be urgent."

Palacios nodded. "I was sick in bed. He got me up. Does that tell you how urgent it is?"

"I won't come."

Palacios was stunned.

"I've been meaning to ask you for a long time, Palacios. Where did you get that red hair?"

Palacios looked momentarily taken aback. "Spanish blood in my family. But, Señora, I think we should go!"

"He's the most important thing in your life, isn't he, Palacios? Mine, too. Funny, don't you think? His former slave and a bastard-born lady doing everything they can to protect the most illustrious man on the continent?"

"Then why won't you come now when he needs you most?"

Manuela stared at Palacios for several seconds then said, "Because I'm not the most important in *his* life."

"No, liberty is more important to him. He would die for it, I am sure." Palacios paused. "My lady, you must not ask him to put you ahead of his cause. He never will; he is not capable."

Manuela smiled sadly. "Believe me, Palacios, I know this better than anybody." She turned away to hide her tears then reached for a pair of thick boots which she pulled over her slippers. "Not very graceful. But at least I'll arrive with dry feet."

Palacios sighed with relief.

"I will get your coat, Señora. We must hurry. I've never seen him look so bad." A distant growl of thunder interrupted their conversation for a moment.

"He's worried," said Manuela. "Things are going badly. Something is going to happen. I hear the rumors from my servants."

"Yes, I am worried too," Palacios said gravely. "He is a sick old man. I wish he would get away. He has achieved our independence from Spain. He's won more glory than any man ever dreamed of."

"As you say, Palacios, he has won his liberty but he cannot avoid fighting. And if he did, these ungrateful countries he created would break up and go to hell. He holds them together. He gives them order and stability. And do they appreciate it? Hell, no!"

"Remember, Señora, you shouldn't talk that way."

"Ah, yes. I must watch my tongue. I mustn't say what I feel about the enemies of the Liberator. I must not have Santander shot in effigy, I must not . . ."

"Come. You're very stubborn. Hasn't your banishment from the palace taught you anything?"

Manuela said quietly, "I love the man. I'll protect him with my life. Even if it makes him mad at me."

The man who opened the bedroom door in the Palace of San Carlos that evening was so slender and lean he seemed ascetic. He looked ill and exhausted. But when he saw Manuela, his tired eyes twinkled back to life and a smile spread across his face. Without a word, they embraced, Bolivar in his robe, nightshirt and slippers, Manuela in her heavy cape and thick boots. After the embrace, Bolivar nodded to Palacios. "Thank you. Good night."

Palacios bowed and closed the door behind him. Inside the room, which was large and comfortably furnished with a bed and a table, two wooden armoires, a couch and several large upholstered chairs, Bolivar and Manuela continued to gaze into each other's eyes. They were both smiling now. Finally, Manuela took off her cape, sat down on the couch and removed her boots, went to an armoire, picked out one of her robes and wrapped it around her. Bolivar's eyes never left her. "You've been away too long."

Manuela rolled her eyes upwards and sighed. She walked to the bed on which Bolivar sat and took both his hands in hers. She sat down beside him, but neither made a move to get under the covers. "I've missed you." Manuela's voice was husky sweet.

Bolivar looked at her longingly.

"Why did you send for me tonight? Why tonight?"

From outside the voice of a sentry challenged a visitor, "Who goes there?" The reply came clearly through the night, "A friend of the Liberator."

Bolivar sighed. "Because I found out tonight that you were right. My enemies plan to kill me."

Manuela said nothing.

"You will be safer here. And I want you beside me if anything happens."

Manuela glanced at the table by the side of the bed. On it were two pistols. On the floor, leaning hilt-up against the table was a sword.

"How do you know they're planning to kill you?"

"An officer named Triana got drunk and told the whole damned barracks my tyranny would be drowned in blood."

Manuela pulled back the covers. She spoke quietly. "Yes. There are several plots, but some are nothing but talk. Maybe Triana was just blustering?"

"He's a friend of that Major Carujo. You know, the short Venezuelan I don't like? If there were a plot, Carujo would be in on it, you can be sure of that."

"Yes, I'm afraid so."

"There is something else that bothers me. Carujo has visited Ramon Guerra several times lately."

"Colonel Guerra? Your chief of staff?" Even Manuela seemed surprised.

"Yes, and the young Frenchman, Auguste Horment. He's been around Carujo a lot, too. I think they are the leaders."

Manuela's eyes widened. "Officers in the army. I knew it . . ."

"Oh, most are loyal to me." Bolivar said the words with satisfaction, even happiness. "Urdaneta, O'Leary, Cordoba, Fergusson . . ."

"I wish Sucre were here."

"I wish he were here too; he's like a son to me. But please, darling Manuela. No more! I need peace. The people called me back to save the republic. That's why I'm the President-Liberator, I rule with the consent of everybody."

Manuela raised her eyebrow and sighed at his naivete. "Come, my Liberator. You've had a long day. It's time to take a bath, then lie down and I'll rub your back."

An hour later, Manuela picked up the candle beside Bolivar's bed. He was snoring softly. She looked into his face and thought, "Palacios is right. My Simon is a worn out old man at forty-five." She shook her head and tiptoed to the large, cushioned chair at the side of the room where she would sleep. From afar the voice of the town watchman called out the hour of midnight. His 'Ave Maria' trailed off into the night. Manuela closed her eyes.

Outside, the sentry called, "Who goes there?"

"Liberty!" came the answer to the sentry's challenge.

Manuela jumped up. She ran to Bolivar. She shook him. She shook him again. He opened his eyes and sat up.

"Hurry! They're here!" she whispered.

He leapt out of bed, grabbed a pistol in one hand and the sword in the other and started towards the door.

"No! Get dressed!"

Bolivar hurriedly pulled on his trousers. Outside, a dog was barking. Bolivar thrust his arms into his shirt. Beyond the door, on the stairs, there came a clash of swords. It was brief, followed by a clatter and a thump. Bolivar was breathing hard. He looked at Manuela. Firing could be heard outside the palace, followed by shouts of "Death to the tyrant! Long live liberty!" Silently, Manuela handed Bolivar her boots. He pulled them on gratefully. "It's lucky you have such small feet," she said. Manuela took Bolivar's arm and guided him to the open window. He sat on the sill while she put her head out to look up and down the street. "It's a long drop. But nobody's out there. If you don't break a leg, you can escape."

Bolivar stood up. He put his arm around Manuela and kissed her. Then, he climbed over the window sill and hung by his hands until he let go and dropped easily into the street. He turned and waved, then disappeared into the night.

Manuela walked to the center of the room. Outside were more shots. More yelling. The voices were desperate now. "If we don't kill Bolivar, we are dead men!" Manuela's face became stern and determined. She picked up Bolivar's pistol and sword. She walked to within a few paces of the door, but as she raised the pistol, her hand trembled. The pistol fell to the floor just as the door crashed open.

A short, unpleasant-looking major with a bushy moustache dashed in, accompanied by a tall, blond man. With them were several soldiers wearing the insignia of the major's regiment. The blond man raced to the open window. The major opened the doors of the closets. A soldier looked under the bed. "Horment! He's not here!" called the major.

"He might have gone out the window!" the other man yelled back. "Come look, Carujo."

The major dashed to Horment's side. "Could be. But I think it's too long a drop."

They both appeared to remember Manuela at the same moment. Both ran to her. Major Carujo shouted, "Where is he?"

"In the council room." Her voice was low and husky.

The two men with their soldiers dashed out of the room together. Their footsteps echoed on the stairs. Manuela continued to stand in the middle of the room holding on to Bolivar's sword.

In a matter of two or three minutes they were back. A soldier snatched the sword from Manuela's hand; another grabbed her around the waist and carried her into the hall outside. "Let me down! I can walk!" she shrieked.

"Let her down," commanded Horment.

The man lowered her to her feet, grabbed her hand and pulled her with him as the group descended the stairs once more. At the foot of the staircase, the officer of the guard lay bleeding from several deep saber slashes. His blood was splattered on the stair and was now pooling on the floor. He groaned. Manuela stopped and tore a piece off her dress to bandage him. The assassins saw what she was doing and shoved her forward. As they did, the wounded officer gasped, "The Liberator?"

Manuela turned and nodded to him. Carujo saw her and hit her on the side of her head so hard she fell to the floor. "Where is he?" he demanded.

"Safe! You murderer!"

Carujo whacked her on the side of her face with the flat of his sword. She toppled onto her side, her ear bleeding from the cut. The soldiers began to close in on her with their knives drawn. "Stop!" shouted Horment. "We are not here to kill women. And we're running out of time. Let's go!"

Carujo kicked Manuela hard. He turned to one of his men. "Take the whore back to the bedroom and keep her there."

The men, including Horment and Carujo, all raced around the ground floor of the palace, searching every possible hiding place. A minute later, a tall, grey-haired gentleman came through the front doors. In his hand was a saber. Coming around a corner, Major Carujo almost ran into him. Carujo jumped back. "Colonel Fergusson!"

"I might have known it would be you, you little jackal!" Fergusson raised his saber and took a step towards Carujo. The major fell back, raised his pistol and shot William Fergusson squarely through the forehead. The colonel went down without a sound, dead before he hit the floor. But the pistol's report echoed through the palace. The acrid smoke hovered in the still air.

"That does it!" yelled Horment, entering the hall from another room. "Let's get out of here!"

Under the Bridge of San Augustin, two men sat in the mud, waist-deep in water. One was Jose Palacios, who had seen his master leave, followed him and was now protecting him. The other was Simon Bolivar. They sat silently. Over the bridge the clattering of horses's hooves cut through the other night sounds. In the distance came shouts, then gunfire. Bolivar shivered slightly in the cold. His voice was hoarse. "Less than a year ago, I was 'the idol of the continent . . . the toast of Europe . . . the spiritual heir of George Washington.' And now, here I am, sitting in the mud, afraid if I poke my head out, somebody will shoot it off."

Palacios patted the Liberator on the back. "You will be all right, General. You always pull through."

"No, Palacios. I'm afraid I will never be the same. Nobody has touched me but I feel I've been stabbed a thousand times. That any of my people would want to kill me has cut more deeply than any sword ever could."

The two men sat in silence.

Presently they heard more shouting in the night. Now the words became distinguishable, "Long live the Liberator! Long live the Liberator!"

"Do you think it's a trick?" asked Bolivar.

Palacios did not answer. He put his finger to his lips, then made a motion for Bolivar to remain where he was. The big man crawled out from under the bridge and disappeared.

A few minutes later, feet splashed in the water by the bridge. Hands reached out to help Bolivar to his feet. He emerged shakily. Soldiers gathered around. At first they were silent, but when they recognized the man they had rescued the cheers burst forth, "Long live Bolivar! Long live the Liberator!" Bolivar lifted his arms above his head to acknowledge the acclamation, and when he did, the shouts became so loud they could have been heard for miles away.

Chapter 56

As soon as Bolivar was safe, he went to the Palace of San Carlos where Manuela Sáenz lay on his large bed, her black hair framing her face against the white pillow. Her bruises showed, despite attempts to cover them with powder. The cut on her ear was hidden by her hair. The door burst open as Bolivar strode into the room. He went to the bed without looking to the right or left. He knelt beside it and took Manuela's hand. Only then did he look up at Dr. Charles Moore, his own physician, who stood by the bed. The English doctor nodded at Bolivar and smiled.

"Then, she'll be all right?" asked Bolivar.

"She has a fever. She's been knocked around a bit, but she'll be fine. When you're thirty, you can survive anything."

For the first time since Bolivar entered the room, Manuela stirred. She looked at him and smiled.

As officials and friends, officers and civilians, came in twos and threes into the room, Bolivar leaned over and kissed Manuela. "Last night, you saved my life." He turned to the crowd of supporters who now filled the room and, pointing to Manuela, declared, "If it were not for this lady, I would not be here among you. She liberated the Liberator." He smiled slightly at his play on words.

Everyone in the room murmured their felicitations. Those who were close enough to the bed patted Manuela on the shoulder. There was relief on all their faces. One leaned over and said, "Thank you, Manuela. We are all in your debt."

The night's rampage left Colonel William Fergusson and several guards at the palace murdered. There was a second group of conspirators abroad that same evening who seized several cannon and attacked the artillery barracks. Colonel James Crofton calmly ordered his sharpshooters to pick off the men manning the guns, then led a group of cannoneers out to turn the cannon around to face away from the barracks and towards the traitors—and after five minutes of firing at the rebels it was all over. The renegade gunners still lay sprawled around their guns, dead, and the attack they were to have initiated broke before Crofton's fire. The conspiracy was over.

Thirty-four-year-old General Francisco de Paula Santander sat in his living room. It was late afternoon on the day after the conspiracy. His friends, including Azuero, and his other political allies were either in custody or under house detention as suspects in the conspiracy. Only three men were with him, and they all wore glum faces. Santander, still a handsome man with brown hair and light skin, seemed sad as well. "In the square this morning, when I went to congratulate the Liberator on his escape from the assassins, he refused to shake my hand. I was the only one he refused. The only one!"

A man named Garcia, shook his head. "And you weren't even in on the plot. You didn't have anything to do with it."

"But I knew there were plots. I was approached. I simply refused to join them. I should have reported them to the government, but there were so many . . ." Still, he *had* refused to listen to the details of their plans. He couldn't assume they were serious. After all, they did a lot of talking that never came to anything . . .

"Everybody knew there were plots," said Garcia. "So why should he pick on you? Besides, you were his closest colleague, his comrade in arms, his vice president. Why, sir, you're a cofounder of the republic! You —"

"That's enough." Santander knew he couldn't rely on past services to spare him from Bolivar's wrath. "I loved and admired Simon Bolivar more than any other man. But he carried his power too far so I tried to stop him. He sees it as betrayal. I am no longer his friend. My neck is already in a noose. One false step and I shall plunge into eternity."

"He still calls you his 'man of laws,'" ventured Garcia.

Santander shook his head. "He used to say it with a smile. Now, he says it with a sneer."

The next day, in the large hall of the palace, Simon Bolivar and General Urdaneta sat together. Other men, both military and civilian, sat with them as a variety of conspirators, all in custody, paraded past, awaiting their fate. Finally, Bolivar rose to his feet. "I have heard enough. Let me speak to the president of the council. Please bring Jose Maria del Castillo to me."

When the little man who presided over the council came to his side, the room fell silent. Bolivar cleared his throat, then announced, "It is obvious from the evidence I have heard already that an extraordinary number of people were implicated in this affair. Too many. Therefore, it must be a popular movement. Under these circumstances, I intend to declare a general amnesty to all those involved."

The other men in the room did not move or speak. They

seemed to have turned to stone. Bolivar continued, "And since it is clear that the people no longer want me to rule, I shall not impose my leadership a moment longer. It is my intention to resign the presidency and leave the country. I shall not stay here. Thank you all for your support and your loyalty."

He bowed and walked slowly out of the room.

Chapter 57

Despite Bolivar's sincere wishes, his friends and colleagues refused to allow him to resign and leave the country with the crisis still unresolved. They pointed out that, if they pardoned such bloody treason and murder, they would only be encouraging the conspirators to stab them as they slept.

He agreed to appoint General Rafael Urdaneta as judge and prosecutor of the conspirators. The ringleaders were hanged, including Horment, Guerra, Triana and the rest.

One evening, after Bolivar had overturned several of Urdaneta's sentences of life imprisonment in favor of hanging, the Liberator brought up the matter of Carujo. "Now, about that little snake, Carujo. He killed my friend Fergusson. He kicked Manuela and hit her with his sword. He led them into the palace, probably killed a couple of my guards."

Urdaneta coughed uneasily. "He is still in prison, of course." He seemed to be thinking hard. "He's been found guilty."

Bolivar waved his hand impatiently.

Urdaneta came closer to Bolivar's chair. He spoke softly. "The little bastard wants to bargain . . ."

"No! He should have been the first one to hang. I'd like to do worse, but I'm not a cruel man. No. He hangs."

Urdaneta's voice fell still lower. "He promises to give evidence that will connect General Santander to the plot. It's something we haven't been able to do."

Bolivar leaned back thoughtfully before asking, "What sort of bargain?"

"His freedom for his testimony—if we convict Santander."

"If we convict him, will Santander hang?"

"If you say he hangs, then he hangs."

Again Bolivar reflected silently. He sat brooding for a long time. "No. No freedom for Carujo. He's a turncoat, anyway. He fought for the Spanish until I beat them. He didn't join us until it was almost over. The man's no good."

"How about his life?" suggested Urdaneta. "Spare him from the rope, but keep him in prison forever?"

"Suppose we don't convict Santander?"

"Then, we'll hang Carujo."

Bolivar shrugged. "Let's try to get Santander. What can we lose? Hanging an insignificant little piss-ant like Carujo means nothing in comparison to nailing the biggest cockroach of them all."

After a heated and controversial trial, even though Carujo's testimony was proven to be absolutely false, and there was no other evidence against him, General Francisco de Paula Santander was found guilty of treason by General Urdaneta and sentenced to hang, also by order of General Urdaneta.

In the large council chamber of the palace, Bolivar sat at the head of a long table. General Urdaneta sat at his right, Del Castillo at the far end facing the Liberator. Other officers and officials filled the remaining chairs. Urdaneta was saying, "We shall execute Santander tomorrow morning," in the same tone of voice he would use to say it might rain.

Bolivar drummed his fingers on the table, looked directly at Del Castillo, and asked, "Do you think Santander should be hanged?"

Del Castillo looked startled. He peered at the Liberator as if he were trying to divine his thoughts so as to give the right answer.

Finally, he replied, "That is the court's decision."

Bolivar nodded. "Yes. I know that. But to hang such a man as General Santander is something we cannot undertake lightly. There will be consequences. What is the council's view?"

Del Castillo spoke hesitantly. "The council has recommended leniency in the case of General Santander. Nothing was proved against him."

General Urdaneta rose and turned to face Del Castillo. The veins stood out on his forehead. He was about to open his mouth, when Bolivar preempted him. "What punishment does the council recommend?"

Urdaneta sat down.

"Perpetual exile." Del Castillo now spoke with more authority. He was expressing the opinion of the council, and he knew now it was also Bolivar's. "That General Santander be banished from the republic, never to return."

Bolivar nodded and said slowly, "The greatest mistake of my life was that I didn't come to terms with Santander. I overrule the court. General Santander's sentence is commuted to perpetual banishment from the country. He is to be escorted to the nearest port and seen safely aboard a vessel. A guard will remain with him until he sails."

Urdaneta looked as if he had been slapped in the face. "And Carujo?"

"Carujo killed my friends," said Bolivar after a short pause. "He hangs."

Chapter 58

Despite the Liberator's restored popularity after the assassination attempt, the plot had destroyed him emotionally. In order to occupy his mind, he resurrected his idea of building a canal through the Isthmus of Panama.

A stronger, more resolute Simon Bolivar stood with General Jose Maria de Cordoba in the study of Bolivar's home in Bogota, with its clutter of books. Bolivar was simply dressed. As usual, the only decoration he wore was George Washington's Yorktown Medal. On a table, lay an open map spread out before them. The two men had obviously been discussing a project. "You see," said Bolivar, "we'll use Panama's Chagres River as far as we can, then start digging."

Cordoba seemed to have caught his commander's enthusiasm. He was nodding his head and smiling. "Yes! Yes! When I first heard about this scheme, I thought it was preposterous. But you have convinced me it can be done. Look, sir. The mountains are low here. And there's a saddle. See?"

"You and I shall work together." Bolivar pointed at the map. "We'll build a canal here at Panama, unite both coasts of Great Colombia and join the whole world by sea."

"We'll need a work force. Remember, Panama's population is small. They're farmers and merchants."

"Yes, exactly," exclaimed Bolivar. "That's the point. We'll employ the soldiers who fought for independence. We have armies of men who need the work."

"Of course. I forgot that you always think of things like that. It will be a magnificent undertaking. I can hardly wait to start."

However, the Liberator's plans to build a Panama Canal had to be deferred.

Marshal Sucre had agreed to be the president of Bolivia for three years. They were not pleasant years. There were insurrections and mutinies. There was a lack of cooperation from the Congress. When the Grand Marshal of Ayacucho was to make his farewell address, there was not a quorum in the Congress. The great Marshal Sucre tossed his farewell speech to the president of the Congress, who stood apologizing to him and said, "Here! You read it when your people find the time to attend to their duties again."

The Bolivian Congress utilized the former Royal Administrative Palace as its meeting place. The president of the Congress stood at the lectern. "I shall read the farewell address of Antonio Jose Sucre de Alcala, Grand Marshal of Ayacucho and President of Bolivia," he intoned. At the mention of that name, the faces in the hall reflected respect—and guilt.

"I would have preferred to deliver my message in person," the man read. "But, unfortunately, a quorum was not present in the Congress. Therefore, by the time this is read, I shall have left you. Three years ago, I promised you I would resign the presidency of this nation today. I am complying with that promise."

The Congress sat still. There was no applause. There was no jubilation. An aura of sadness seemed to hang over the assembly like an Andean mist.

"The constitution," continued the presiding member, reading Sucre's words, "states that I cannot be held responsible for any acts of my government. I renounce this privilege. I ask, instead, that my conduct as president be examined, and if I have committed any act in violation of the law I shall return from Quito and submit myself to the sentence of this House.

"I demand this as my right, because I solemnly declare that I, myself, ruled during my administration. Whatever good or evil has been done, I am solely responsible for it, and I should be held accountable for it. Farewell."

Chapter 59

Simon Bolivar had to abandon his plans for a Panama Canal because his dream of a united South America was turning into a nightmare. Revolts and counter-revolts split the nations asunder. Marshall La Mar, now president of Peru, marched into Guayaquil with the purpose of annexing that city to Peru. With an army one third as large as La Mar's, Marshall Sucre defeated him decisively at the battle of Tarqui, while General Cordoba defeated the brigand, Obando, but failed to capture him.

Bolivar had gone south to join his two most brilliant generals when he learned of their victories.

Inside the large, solid house, Marshal Sucre sat in the living room while Bolivar slept in his bedroom. He didn't seem to mind waiting for his commander in chief to finish his siesta. Sucre was enjoying the sheer beauty of the room—furnished in mahogany and set off by white walls and black iron sconces—when General Cordoba entered.

"I hear you were against this war with Peru, too," Cordoba said to him.

Sucre nodded. "Yes. It was senseless. The Peruvians deposed La Mar, as we knew they would even before Tarqui. The new president of Peru has sent word to the Liberator that their country would never forget the extraordinary services Colombia rendered her. They've withdrawn from Guayaquil. All this would have happened without our having to fight our old friends."

"And we could have finished off Obando. The Liberator doesn't listen to me anymore. If he had, he might have spared himself the ordeal of September 25th. I told him about Manuela and Crofton shooting Santander in effigy, and the Liberator promised me he'd do something about it. Do you know what he did?"

Sucre shook his head.

"He continued living with Manuela, and he promoted Crofton." There was a bitterness in Cordoba's voice.

"Take it easy, Jose Maria. And be careful, the Liberator isn't well, you know."

"Do you know what else he did?" Cordoba was obviously warming up to his subject. "He let that damned Obando get away without even a slap on the wrist. And the man is a scoundrel, a brigand, a cruel murderer, a bloodthirsty renegade. Now he's walking around free as air, while the idealistic conspirators of Bogota paid with their lives. It's unfair."

Sucre tried to placate his friend. "The Liberator didn't know I'd already beaten the Peruvians at Tarqui. He didn't know the situation and wanted to get south as fast as possible, and it was quicker to let Obando go than to take the time to fight him."

"Bolivar was my idol."

"What do you mean, was?"

"I'm getting disillusioned," said Cordoba.

Sucre stood up. Cordoba turned around. Simon Bolivar was emerging from his room, stretching as he approached. He wore a pair of old trousers and a woolen poncho. He embraced both of his uniformed generals.

"I see you promoted O'Leary to general on the battlefield of Tarqui," he said to Sucre. "I'm glad. It's about time, but I didn't want to do it. It would look too much like favoritism."

"He deserved it," said Sucre. "At Tarqui, O'Leary led the charge that broke the enemy line. He's good. I need him and I'd like to keep him as my second in command."

Bolivar nodded. "Fine for now. But I shall want him back."

Cordoba handed Bolivar an envelope and said, "I came to deliver this, sir. Now I have to get back to my troops. They're still camped outside Pasto."

Examining the envelope, Bolivar asked, "What's in this? And

when did we start using our best generals as messengers?"

Even Cordoba had to smile. "It's from me, sir. It's my request for a higher command, maybe chief of staff . . ."

"I'll consider it," said Bolivar, waving his hand as an indication Cordoba was free to return to his army. Cordoba smiled and waved back as he left.

Sucre looked puzzled. "Sir, I thought you might have already decided to make General Cordoba your next chief of staff. He deserves the promotion more than anybody else. His heroism in battle after battle entitles him to . . ."

"No. I've already appointed another chief of staff. Besides, I'm not sure of his loyalty anymore. I've heard that he's been complaining about my being dictatorial, saying that I want to be king and a lot of other nonsense. Let him cool off a little bit. And don't talk to him so much from now on."

"Yes, sir." Sucre accepted Bolivar's decision with obvious reluctance. Like many others, he realized that since the September 25th plot the Liberator no longer trusted people the way he used to. "Are you going to appoint him to higher command?"

Bolivar didn't reply.

"But, sir, you must do something for him. Promote him, somehow."

"I'll bring him into the cabinet."

"That's even better!"

"I'll make him Secretary of the Navy."

"Uh, sir, we don't have a navy."

"Exactly," croaked Bolivar. Then, he grinned.

"Please be serious, sir. I think the man's getting fed up with your treatment of him. He's rash and hot-headed. I think we ought to try to soothe his vanity a little bit, somehow . . ."

"You're right," agreed Bolivar. "But I don't trust him. He's conceited." Bolivar had been persuaded by Manuela to consider Cordoba a dangerous rival. "Let's make him Ambassador to some European country. Will that placate him, do you think?"

Sucre seemed to relax a little. Perhaps Bolivar would listen to reason, after all. "If it's not too late."

"Then, let's pacify him quickly before he does something rash. In the meantime, I've written to one of his colonels asking him to keep an eye on General Cordoba."

"Good God! I hope he never finds out you did that."

Chapter 60

A week later, in the camp outside of Pasto, twenty-five-year-old General Jose Maria Cordoba stood shaving in front of a mirror propped in the crotch of a tree. His shirt was off, and he was humming to himself. Squatting on the ground beside him, his orderly was busy polishing boots. Cordoba stood back to look at himself in the mirror and run his hand over his now smooth face. In high spirits, he said aloud to the mirror, "Just look. The face of Jose Maria Cordoba, the handsomest, bravest, youngest general in the entire army." Getting into the spirit of his self-admiration, which was only partly in jest, he turned to the orderly. "Tell me, Pepe. What do I lack? Is there anything at all I need that I haven't already got?"

"Yes, sir. A little bit of modesty."

Cordoba was about to put the man in his place, when he was interrupted by the approach of a Colonel Jimenez, one of his commanders. Jimenez's face was flat and his hair was stiff, straight and jet black, displaying his Indian extraction. He looked worried. "Sir, I have a letter. A man gave it to me."

Cordoba looked at the envelope. "It's addressed to you, personally. Who's it from?"

Jimenez hunched his shoulders. He appeared sullen.

"You can't read, can you?"

Jimenez shook his head. "I was commissioned on the battlefield after all my officers got killed. I got promoted for fighting good and killing lots of Spaniards, but I never learned to read."

Cordoba smiled. "I'll be glad to read it for you."

The man handed the letter to Cordoba, seeming relieved.

"It's from the Liberator, himself!" Cordoba said with surprise. "Why in the name of hell would the Liberator write to a colonel in my army? I'm in command here."

Jimenez shrugged again. "He knows me. From the old days of Boyacá and Carabobo. I told him I was his most loyal soldier, and he said he knew he could always count on me."

By now Cordoba was red in the face. "You want to hear what he says? He sends you greetings and all that. Then, he tells you to use your sword against any attempt at rebellion by anybody of any rank! You hear that? Do you know what he means? He means me!"

Cordoba was now shaking with fury. The blood vessels pulsed on his forehead. His face was the color of a bull fighter's cape. "You know why he said that?" he shouted at the hapless Jimenez. "Because he knows I won't submit to an emperor named Bolivar. That's why! Because Bolivar wants to make our republican form of government so contemptible all Colombians will fall on their knees to beg Bolivar to place a crown on his own head. That's why!"

"Yes, sir. I think I'll go now, sir."

A glint came into Cordoba's eyes. "Bolivar is nothing but a sick old man. The country needs a young leader, a strong, courageous leader. A popular hero." He looked straight into Jimenez's eyes and asked, "Why not me?"

General Rafael Urdaneta and his plump, attractive wife, Dolores, were entertaining Manuela Sáenz at dinner. Dolores Urdaneta seemed uncomfortable having Manuela in the house, but she knew that her husband's loyalty to Bolivar obliged her to be nice to Manuela. Like every other citizen of Bogota, Dolores loathed Manuela for flaunting her extravagance during an era of extreme hardship and poverty. Manuela was dressed by the best

Parisian couturiers and wore the most expensive perfumes. Her Quito lisp made it clear she was not from Colombia or Venezuela. She was an outsider.

As they were chatting at the dinner table, a servant entered the dining room and announced, "There's a messenger at the door, sir. He says it's urgent."

Urdaneta got up and was about to excuse himself when Manuela stopped him. "No, Rafael. Bring him in. We all want to know what's happening."

Dolores looked up at Manuela, shocked by her boldness. How dare she contradict her Rafael! And in his own house!

Hesitantly, General Urdaneta nodded to the servant, who bowed his understanding and went to get the messenger.

Wet and muddy, the man stood in the door, reluctant to enter the room.

"What news?" Urdaneta was outwardly calm.

"General Cordoba has revolted, sir. He's seized the barracks at Medellin. He's got an army supporting him already, and a lot of the local people are rallying to him. It looks like a full scale rebellion, sir."

Urdaneta looked appalled and anxious. His mind was working furiously.

"I knew it!" cried Manuela. "I told the Liberator not to trust that man."

Again, Dolores seemed stunned by Manuela's outburst. She herself would not have dared to speak.

Urdaneta waved at the messenger. "Thank you very much, son. Now, you go into the kitchen and they'll give you a good, hot meal."

The man grinned and left with the servant.

"How serious is it?" Manuela asked.

"Not terribly—if we stop it quickly. I think the young hotspur mistakes popular acclaim for popular will—but I wish it hadn't come to this. Everybody likes young Cordoba. He's one of the most popular men in the country. Dammit! Why did he have to do this? It's madness."

"You can't let him get away with it." Manuela spoke firmly

with a hint of malice. "If you do, then, everybody will want to rebel. He has to be put down right away. And executed. You hear me, Rafael? Executed!"

Both Urdanetas stared at Manuela in dismay. Such a hard, vengeful stance did not become a lady. Urdaneta tried to dissuade her gently. "I think we should try to placate him, somehow. He's too good to shoot."

"No! He dies." Manuela's words were practically hissed.

Dolores Urdaneta rose from the table, glared at Manuela, then glided out of the room without a word.

General Daniel O'Leary stood before General Urdaneta in the president's office in the palace. "I'm glad you're back from the campaign," said Urdaneta. "As you know, the Liberator left me in charge while he's in the south."

O'Leary nodded.

"You know about Cordoba's revolt?"

"Everybody does."

"You're our best officer. That's the reason I'm sending you to put down the uprising."

O'Leary started to object.

"I'm going to have to order you to go." O'Leary could tell that Urdaneta was not going to bend on this matter. "You'll take nine hundred selected men from the British Legion. I don't want to risk Colombian troops in a situation like this. They might defect. The British won't."

"Can't we try to bring Cordoba around? The Liberator planned to name him Ambassador, you know. To some European country. Holland, I think . . ."

"We've tried everything. He won't listen. He's headstrong and thinks he has the support of everybody. Also, he refuses to trust the Liberator."

"Maybe I can take him prisoner."

Urdaneta thought over that prospect, then shook his head. "No. That would be worse, because, you see, Manuela Sáenz hates Cordoba as passionately as she loves Bolivar. She considers him a threat to the Liberator and has decided he must be executed if he's captured."

"And the Liberator will do whatever she wants. You know his execution would be the worst thing that could happen. That could easily cause the government to fall. To say nothing of the humiliation and degradation of one of the finest heroes of the republic. No. It's unthinkable."

Urdaneta observed sadly, "Rebellion is not to be tolerated either. We can't allow it to continue."

The two men were silent. When O'Leary looked into Urdaneta's eyes they reflected his own comprehension. O'Leary summed up their unspoken understanding. "Cordoba has to be killed in battle, fighting against the republic. God! I hate him for making us do this."

The smoke still hung over the field. There were bodies on the small ridge, a few scattered muskets strewn along the rise, a half dozen wounded soldiers returning from the scarred slope. General O'Leary sat on his black gelding surveying the scene. A young officer rode up and saluted. "Resistance has ceased, sir. The rebels who stayed to fight have all been killed or rounded up. A couple of dozen might have gotten away. No more."

"General Cordoba?"

The man shook his head. "Don't know, sir. He's not a prisoner. If he's not among the dead, then he got away."

O'Leary dismissed the officer and turned to watch as his second in command, Colonel Rupert Hand, came marching down the ridge with a small contingent of men at his back. He was smiling, obviously happy the battle was over.

O'Leary saw him and called out, "Hand! Come over here, will you?"

Still smiling, Hand approached O'Leary and saluted smartly.

O'Leary was not smiling. He looked extremely unhappy. "Rupert, I think Cordoba got away. Take those men there with you and find him. And, when you find him, kill him."

Hand drew back in revulsion. "I simply can't do that. I know General Cordoba. I've drunk with him. I've . . ."

"Rupert!" interrupted O'Leary. "You're the only officer available right now—and, incidently, the only one I can trust. This has to be done and done quickly. There's a good reason for it. Do you understand?"

Hand didn't reply.

"Rupert, I'm ordering you to go find Cordoba and kill him." With that, O'Leary wheeled his horse and loped off toward the ridge.

Rupert Hand had proceeded cautiously for the last half hour, noting each small drop of blood that had clung to the leaves and the grass in front of him. His men were fanned out, American Indian style rather than in the formal European manner of straight-lined ranks. Before them was a native hut. Motioning his men to keep back, Hand followed the blood drops to the door, which was open. He was breathing hard. From his belt, he drew a pistol, made sure it was loaded and cocked, then entered the hut. In a corner, a man gasped. Then, he recognized the intruder. "Thank God it's you, Hand. I'm wounded badly."

Hand took two steps nearer the wounded Cordoba. He held his pistol out of sight in his hand alongside his leg until he was close enough to look into Cordoba's eyes. He could see both hope and relief reflected in them. It nearly deterred him, but he steeled himself. In one smooth motion, he pointed the pistol at Cordoba's chest and pulled the trigger, hoping Cordoba wouldn't realize what was happening. The report startled him; the flash of the powder blinded him. He turned and staggered out of the hut, leaving Cordoba dead on the floor.

Outside in the sunlight, Rupert Hand was white as chalk. He tossed his pistol into the dirt in a motion that was at once defiant and contemptuous. He unbuckled his belt and let it drop to the ground; he took off his uniform jacket and tossed it into the bushes. His eyes were slightly glazed as he walked away. His men automatically fell in behind him in marching order. Two of them picked up his discarded gear. One carefully wiped the pistol and stuck it in his belt.

* * *

The debris from the battle had been removed when Rupert Hand arrived back on the field with his men an hour later. He croaked to the first officer he met, "General O'Leary?"

The man pointed to the ridge. Hand nodded and headed in that direction. O'Leary, still mounted, was riding back and forth, directing the clean up operations and congratulating his troops on their performance. He saw Rupert Hand and rode over to him. Hand, who still looked white and sick to his stomach, saluted feebly. "I did what you ordered me to do."

"I knew you would." O'Leary's voice was flat. There was no trace of approval or disapproval in it.

"What I did made me sick. I'm through. I'm resigning my commission and going home."

O'Leary looked sympathetic but said nothing.

"You should leave, too. Our glorious revolution has turned into a monster that devours its young."

"My devotion to the Liberator won't let me." O'Leary spoke with infinite sadness. "I think things are going to get worse, and he's going to need me."

"Goodbye, Danny," said Rupert Hand. He turned his back and walked slowly down the ridge.

Chapter 61

Grand Marshal Sucre of Ayacucho, alarmed at the reports he had heard of the Liberator's deteriorating health and growing disillusionment with the countries he had liberated, went north to Bogota. When he first saw the Liberator, he was dismayed.

"The Liberator has been ill," Manuela explained, "and things are not going well. Tell me. How is your family? Mariana and your little daughter?"

"Wonderful," he said. "I think sometimes I would not have fought so hard if it were not for my daughter. She is the one who will profit from our efforts." Sucre was very good at hiding his feelings. He knew his beloved wife, Mariana, was having an affair with his aide-de-camp and it had broken his heart. He could never be like his idol and mentor, the great Liberator, and take one lover after another. The gallant Marshal Sucre was a loyal, one-woman man. Now all his love was focused on his daughter. Sucre's will, which he wrote before departing Quito for Bogota, declared, "At this moment my wife, Mariana, is not pregnant, and I therefore leave all of my estate to my daughter, Teresa . . ."

Bolivar began coughing uncontrollably. When he put his handkerchief to his mouth, it came away red. Afterwards, his breath was labored, yet he held up a hand, and spoke in a low, hoarse voice. "While I was away, Urdaneta and the cabinet handled things badly, especially in the matter of Cordoba's revolt. The people hold me responsible. They haven't told me yet, but I know the feeling is that as long as I remain here, there will never be peace or stability."

Sucre paused before bringing up the next bit of bad news. "Yes, I know. I just heard about Venezuela."

"Paez has decided he's strong enough and I'm weak enough for him to separate Venezuela from Great Colombia. How do you like that?"

Sucre shook his head. "I'm afraid, sir, you should have finished him off when he seceded from Colombia three years ago. You trusted a man you knew was untrustworthy. Now, I think it's too late."

Bolivar seemed extraordinarily depressed by Sucre's words. "I am a failure." Dejection had replaced his fit of coughing. "Independence is the *only* good thing we have achieved at the expense of everything else." A look of complete despair swept across Bolivar's face. "That was my mission."

Manuela remained silent. Sucre said, "I'm afraid you're becoming bitter."

"Why wouldn't I be? I wanted to lead our America out of the

war as we entered it—united. All my political strategy was directed towards that end. But who can say, honestly, that one person is better off than he was before?"

Warming to his subject, Bolivar continued, "There is neither faith nor truth in America, whether it be among men or among nations. Treaties are mere scraps of paper, constitutions are books, elections are battles, freedom is anarchy, and life is torture. This is our situation, and if we do not change it, it were better that we should die."

Sucre and Manuela remained silent. After those dismal words, there was nothing they could say. Seeing the Grand Marshal to the door, Manuela said, "He has never been the same since the September plot. It broke his will. It disillusioned him with his people and his cause."

In his room alone, Bolivar put his head in his hands and fought back tears.

I am going mad, he thought.

The next day, Sucre tentatively entered Bolivar's study. The Liberator was sitting at his small desk, dressed in a faded tunic and a pair of old army trousers.

As usual, his only decoration was George Washington's Yorktown medal, which he wore all the time now.

He put his hand to his head, a gesture he was making more frequently as his illness worsened. "Nobody understands the importance of uniting these countries." Then, slamming his fist onto the desk, he suddenly stopped and dropped his head into his hands. "It's too much. Too much. How long can I take this? I've suffered enough. I can't go on."

Sucre looked alarmed. Manuela appeared at the door having heard the outburst. She quickly escorted Sucre to the door. When he turned to bid her farewell, his eyes were moist. The two stood silently. Then Sucre said, "I think the Liberator is disintegrating even more rapidly than the republic."

Several days later, a slightly restored Simon Bolivar walked into the living room of La Quinta. Waiting for him were several men he felt he could trust, including General Urdaneta and Pedro Herran, the provincial governor of Bogota. Marshal Sucre was in the city, presiding over Congress. The men stood as he entered. "Sit down. Please sit down, gentlemen."

As the other gentlemen sat, the Liberator continued to stand at his desk as he spoke. "As you know, one of our regiments on the coast deserted to Paez in Venezuela."

The men had all heard the news.

"Something has to be done. Otherwise the country will break up completely. Therefore, I propose to assume authority and declare war on the Venezuelan secessionists . . ."

"No. No," said Herran. "Such a war would be impossible. The separation of Venezuela is a fact. It must be accepted as such."

Bolivar turned to Urdaneta. "What do you say?"

"The separation of Venezuela was consummated when you pardoned Paez in 1827. It was you, yourself, who sounded the death knell of the republic."

Bolivar looked as if he had just been hit by a musket ball.

"And I have a message here from Jose Maria del Castillo," continued Urdaneta. "He couldn't come today, but he wants to advise you to renounce power forever. The separation from Venezuela is a fact. Any war to reunite us would be most unpopular. An independent government must be constituted. A government without Bolivar."

The Liberator shouted. "I won't go! I'm staying in spite of you all. You want me to desert the republic when she needs me the most. You're nothing but a bunch of cowards and hypocrites. To hell with you all." He had to pause to catch his breath, then stormed on. "Christ! I should have accepted the dictatorship when it was offered me. Then I wouldn't be hamstrung by these damned constitutions and laws. I could have ruled by decree. I could have united these insane, chaotic countries. The army would have enforced my will."

Suddenly he stopped, his shoulders drooping. The others looked on uncomfortably.

"No! I refuse to force my will on the people. It is against everything I believe, everything I have fought for. Perhaps this is my fault. Perhaps I never properly showed the people the virtues of democracy."

Then, with a sigh, he headed for the door. "It doesn't matter," he muttered over his shoulder. "My name already belongs to history, and there I shall eventually find justice."

As the visitors gathered their cloaks and greatcoats, Herran whispered to Urdaneta, "You know, he's dying to go to war again. He thinks he can regain his glory only on the battlefield."

Urdaneta shook his head. "I feel we've been in attendance at the agony of a great man."

"We've been trying to get him to see the United States as a threat, to divert his attention in that direction."

Urdaneta shook his head. "You'll never be able to do that. You know how he feels about the United States."

"I know. And he's still able to think clearly sometimes. He pointed out to us that the United States is expanding *westward*. They've made no move to the south, not even to Mexico or the islands in the Caribbean. They have no army to speak of, and they're a long way away from us."

Herran nodded. "You heard what the Liberator said the other day. 'This country will pass through all forms of government until the day dawns when the Anglo-Saxon race invades Hispano-America in a democratic fashion and one immense nation is formed . . .' He sees the United States as our eventual patron and benefactor. That's his problem. He's trying to emulate them."

Chapter 62

The Colombian Congress debated heatedly whether to confirm Simon Bolivar as president or to elect a successor. Many considered Bolivar to be a disrupting and destabilizing influence on the republic.

After several weeks of suspense, Manuela came running into the Liberator's study, short of breath. "They're here."

Bolivar stood up and made his way wearily into the living room. He greeted the three men who had just arrived. "I can see by your faces you bring me bad news."

General Caicedo, who was presiding over the Congress as its acting president, nodded. "Yes, sir. The Congress unanimously agreed that for the sake of peace and the integrity of Colombia it is advisable that the Liberator should not be reelected president."

Bolivar stood still. He was obviously hurt.

Caicedo was a friend of both Bolivar and of Sucre, who had appointed him to preside, with the Liberator's approval. He was a tall, good-looking man with a receding chin and balding, light-brown hair. He wrung his hands and fidgeted, perceptibly embarrassed. "We are also to inform you that your continued residence in the capital of Colombia constitutes a threat to domestic peace."

Bolivar grew red. He stamped his foot defiantly and picked up an ink bottle and hurled it across the room. Then, a fit of coughing interrupted his tantrum. He wiped the bloody froth from his mouth after the coughing subsided. Composed and calmer, he asked, "What will my position be, now that I have declined the presidency?"

"Sir, you will *always* be the first citizen of Colombia."

Thus it was that on May 8, 1830, the great Bolivar rode into the mist on his way to exile, after embracing his friends of the Diplomatic Corps, the Cabinet and the leading citizens of Colombia. At the end of a long day's ride from Bogota, an exhausted Liberator arrived at a large farmhouse by the river, which had been appropriately prepared ahead of time for him and his party. Even though he appeared weary, he asked for paper and quills. Seated at the large kitchen table, he wrote:

My dearest Manuela,

I fear you still don't realize why I couldn't take you with me. I explained to you once, and I do so again: I do not know where I am going. Perhaps France. Maybe Jamaica.

My course isn't clear to me yet. But, as soon as I know, I shall send for you. Anyway, my beloved, I am happy to tell you I feel well, but I am just as filled with grief over our separation as you are. I love you very much and shall love you much more if you will now be more reasonable than ever before. Be careful what you do, or you may ruin yourself, and that means both of us. I am always your devoted lover,
Bolivar.

He re-read the letter, then sealed it. From his shirt pocket, he took out another letter and unfolded it. Bolivar sighed. "Ah, Sucre, I just couldn't bring myself to say goodbye to you."

He read Sucre's letter:

I think you purposely misinformed me of the hour of your departure because when I came to your house to accompany you, you had already left. Perhaps this was just as well, since I was spared the pain of a bitter farewell. In this hour, with my heart so oppressed, I don't know what to say to you. Words cannot express the feelings of my soul, but you know my emotions, for you have known me a long time. And you know it was not your power that inspired the warmest feelings in me, but your friendship. I shall always keep that friendship no matter what destiny awaits us, and I flatter myself that you will always keep the opinion you have had of me. I shall try to be worthy of it under every circumstance. Adieu, my General. Receive as a token of friendship these tears shed for your absence. Be happy wherever you may be, and wherever you are you may count on your faithful and devoted,
Sucre

On the second day of his ride to exile, The Liberator seemed more weary than usual. He slouched in his saddle, his head bowed down to his chest like a dead flower on its stem. Slowly he reined in his mare and dismounted. Beside the dirt road was a bench built by the Spanish Imperial Government years ago for weary travelers to rest safely on their journey. The bench wood was now discolored and termite infested.

Beside the dilapidated remains was a tree stump. It hadn't been a large tree, but, then, the Liberator was not a large man. He sat on it easily and turned his rheumy eyes to Belford Wilson.

When at last he spoke, his voice was weak from the day's ride. "Antonio Paez was right, you know. He told me: 'We are *not* the country of George Washington.'"

Belford remained silent. He knew no answer was expected of him.

"I was wrong, Wilson. I attempted to impose an unsuitable system of government on the nations I liberated. I reasoned, my friend, that democracy was the ideal political structure on which to build our continent." Bolivar shook his head sadly. "And it has brought me to this. South America has no more understanding of democracy than a peon has of Socrates."

"What would you have, then, sir? I know you detest a monarchy."

The Liberator sat silently for a while. Shaking his head slowly, he said, "I'm not certain. For us, the result of democracy is chaos. That is sure. That has been proven. What form of government would you have chosen?"

The question took Wilson by surprise. Finally, he said, "I'm from England. Our system works. As you know, sir, we're a constitutional monarchy."

Bolivar waved his hand. "I know. I know. The Spanish system worked, too. The colonial system worked. It worked for 300 years, but we've destroyed all that. No. *I* have destroyed all that. I attempted to replace it with democracy. And it hasn't worked."

"Then, sir, if I may ask, why did you decide to impose democracy on your people? After Ayacucho, sir, you were God. The people worshiped you. They would have accepted anything you proposed to them. Some countries wanted you to be their dictator. Some wanted you to be their king. You refused all of those offers."

Again the Liberator nodded. "I had no personal ambition for myself. I wanted to leave a self-governing continent, free and democratic."

"You didn't realize such a scheme was impractical?"

"No. I suppose, Belford, that I didn't really understand my countrymen. I never thought they'd resist democracy with such iron tenacity. They don't want it!"

"All they wanted was *you*. They wanted you to govern them without any hindrance. It would have worked, you know. Nobody would have dared raise a hand against you."

The two men remained silent.

Bolivar then said slowly, "What is different between us and the colonies to the north? The North American colonies revolted from England and established a democracy which still functions . . ." Seeing that Wilson was about to interrupt him, Bolivar raised his hand to stop him. "In the United States, they have one language, one people. They are an agrarian society. In South America, we have one language, one people, and we are an agrarian society. What makes us different, Belford?"

Without hesitation, Wilson replied, "England."

Bolivar nodded. "Yes. The democratic heritage of England. That was all we lacked." He smiled. "That was all."

Chapter 63

On May 25, Bolivar and his little group of retainers arrived in the old city of Cartagena, with its narrow streets and ancient Spanish buildings. The plaza was still dominated by the old palace of the Inquisition and the Municipal Building. It was an important port, even though it was small. Spanish-built forts guarded the harbor. A large, open house had been prepared for the distinguished visitors, and the Liberator immediately entered it and sat down on a small chair. The weather was hot, and he was sweating. Nevertheless, he seemed relaxed and happy to be in the port from which he intended to sail away forever. Several curious young boys were bold enough to invade the verandas and porches that surrounded the house and peer in through the open door and the windows to see what the most famous man in the world looked like.

Their heads suddenly disappeared, as a squadron of horsemen galloped up. A portly, florid officer dismounted and entered the house. "I am sorry not to have been at the gates to meet you, but I didn't know when you were arriving."

"No. No, Mariano," said Bolivar as the two men embraced. "Let me introduce you to my staff. Gentlemen, this is the famous General Mariano Montilla. He's the man who took Cartagena, and he's the man who governs it. I've known him since we were children." He then presented each one to General Montilla.

"I'd like to have you dine with me tonight," said Montilla after the introductions.

"Thank you, not tonight. I'm not up to it. I'm not well, Mariano."

"I want to beg you not to leave us. The nation needs you more than ever."

"No. No. I come here as an exile. From Cartagena, I'm taking a ship to Europe."

"You can't do that." Montilla was pleading now. "Besides, if you go abroad, you will be forced to live like a beggar. You won't be able to support yourself, Simon . . ."

Bolivar smiled at his old friend. "If I don't die on the voyage, I doubt if the British will let me die of starvation. Besides, Colombia wants it this way."

Antonio Jose Sucre de Alcala, Grand Marshal of Ayacucho, was the undisputed heir of Bolivar. He presented the only major threat to the anti-Bolivarian factions of the region. Sucre was the hero of his soldiers and of all who were loyal to the Liberator. Always a modest man, he had no fancy for politics and hated the arguing and bickering of the legislators. He considered himself no more important than any other private citizen and he wanted to go home to his little daughter in Quito where he intended to live quietly and wait to see what the future held.

In Bogota, Sucre mounted his horse. Across the saddle were his bedroll and several pouches containing provisions for his trip to Quito. General Caicedo and General Urdaneta stood by with several other officers. Urdaneta stepped forward. "I urge you again: Please take a squadron of Hussars with you. These are wicked times. It's too dangerous for you to travel alone."

Caicedo suggested, "Go by sea. It's safer. You read the article in their scandal sheet. It said, 'Perhaps Obando will do to Sucre what we should have done to Bolivar.' I think it's plain. They are clearly plotting to kill you."

Sucre shook his head. "Who would want to do that? I'm just another retired soldier going home to his family."

"For the sake of God and the country, please be careful," urged Urdaneta.

Sucre waved and spurred his horse forward.

The trail was narrow and winding. Sucre was never a good rider; he swayed awkwardly in the saddle. He made an infinitely lonely figure as his horse felt its way slowly down the path, which was slippery from the rains. Except for the swishing of a small mountain stream, all was quiet. Sucre began to hum a little tune. From behind him, two shots shattered the stillness. The first passed harmlessly over him, the second hit him squarely in the back of his head. Marshal Sucre was dead before he hit the ground.

"And he lay in the mud for two days before anybody found him," said Belford Wilson.

Bolivar wiped the tears from his cheeks. "My God! They have shed the blood of Abel. How can I stay in such a country, where the most famous generals are cruelly and barbarously murdered, the very men to whom America owes her freedom! I cannot live among such murderers and rebels. I have no honor among such scum and no tranquility under their threats."

* * *

It was hot, and the dust hung over the streets of Cartagena, kicked up by the horses of the dispatch riders who came loping into town. The stables were full, and horses were tethered in the streets. The smell of their manure permeated the old city. All activity centered on the house in which the great Bolivar waited for a suitable ship to carry him off to Europe.

The house had a wide porch in the back, and Fernando Bolivar, the Liberator's bright young nephew, sat on a swing there, looking pensive and downcast. Belford Wilson came through the doorway from the house. He was sweating, even in his shirt sleeves. "Damn! It's hot here on the coast."

"Tell me again what the hell's going on," said Fernando.

"Ask the Liberator. You're his favorite nephew."

"He's busy right now. And he's frail. When he's not reading the stuff he's getting from Bogota, he has to rest, and I'm damned if I want to disturb him."

Wilson smiled. "You're right. There have been too damn many dispatch riders coming into town, and each one brings a pouch full of letters. There's not any way he can read them all, even if he were to read night and day. Even I have lost my good humor."

He cocked his head. "Wait a second. I hear a whole group of riders coming in together. Fernando, it's an eight-day trip from Bogota. There's at least ten of them coming at once and they're all gentlemen. So it must be something big."

In the wide-open living room of Bolivar's Cartagena residence, a delegation of well-dressed gentlemen stood facing the Liberator. Curious residents of the city peered in through every open window and door. As they watched, a tall man with a bald head and thick moustaches stepped forward. He seemed to know the Liberator and he appeared pleased as he said, "Your Excellency, we come from General Urdaneta to inform you that Don Joaquin Mosquera has resigned the presidency, and General Urdaneta is now holding the office only until you can return to resume your duties as president of the republic." At this point the man smiled and bowed from the waist.

Bolivar's face remained stern and expressionless. Standing beside him, Fernando Bolivar and Belford Wilson also smiled broadly, as did the men who stood behind them.

The spokesman continued, "The ambassadors of Great Britain, the United States and Brazil have declared publicly that you alone can save Colombia."

At this, Bolivar nodded curtly.

But the spokesman hadn't finished. Bowing from the waist once more, he proceeded, "The political and military leaders of the country have appointed you chief of the army and granted you complete freedom to take any measures necessary to maintain order in the republic."

The man stepped back, and the delegation stood expectantly, clearly waiting for an acceptance speech from the Liberator. Instead, he sat still, looking wan and tired. Finally he said in a weak and tremulous voice, "Gentlemen, I am old, sick, disappointed, grieved, calumniated and badly paid."

The delegation stood as if they were in shock. The Liberator looked solemn and unbending. "Believe me, I have never looked upon insurrections with friendly eyes, and during these last days I have regretted even those we undertook against the Spaniards."

The delegation gasped.

"All my reasoning leads me to one conclusion. I see no hope of saving the nation."

After the commission from Bogota had left and the curious had been cleared off the verandas, Bolivar asked for a cup of hot tea. He left his easy chair, walked over to the table and sat down on the straight backed chair next to it. The table was piled high with letters and petitions begging the Liberator to return to power. He pushed them aside to make room for the tea.

"It's been quite a day, hasn't it?" said Belford Wilson, who stood nearby.

Bolivar nodded wearily.

Fernando returned from the kitchen with the cup of hot tea for his uncle. He put it down on the table. "You certainly didn't give them any encouragement this time, did you? Then, when you smiled and said farewell to them so graciously, they could only mumble their goodbyes, as if they were in a trance. Do you want me to try to call them back?"

Bolivar leaned his elbow on the table to prop up his head. "No. Everything is lost forever. America is ungovernable." He drank some tea before continuing, "What benefit have we gotten from twenty years of war and revolution?" He answered his own question: "None! We have ploughed the sea!"

Chapter 64

By now, Bolivar was suffering from advanced tuberculosis complicated by cardiac disturbance. Describing him as near death, Bolivar's doctor refused to let him travel to Europe, knowing he would never survive the journey. Instead, Don Joaquin de Mier, an ardent Spanish loyalist but also an admirer of the Liberator, offered Bolivar his sugar plantation in Santa Marta as a peaceful place to live out his days.

On December 7, Jose Palacios lifted the emaciated, blanket-wrapped, but smiling Bolivar into the oxcart lined with straw. As the cart moved off, Palacios turned to Fernando Bolivar and said, with tears streaming down his cheeks, "Just think, Don Fernando. This might be his last journey alive, and he's making it in an oxcart!"

Fernando's words came hard. "You must always remember him in the saddle, Jose, always leading the charge . . ." His voice broke, and he put his head against Palacios's chest and sobbed.

General Laurencio Silva, who was now married to one of Bolivar's nieces, had come to join the group attending the Liberator.

"What does Don Joaquin call this place?" he asked Fernando.

"San Pedro Alejandrino," replied Fernando. "Isn't it a gem."

The men looked out over the palm trees swaying in the breeze, at the garden and the magnificent mountains rising out of the lush green of the cane fields. Behind them, the sparkling blue Caribbean sea rolled noisily onto the sand.

"It's a perfect spot," Belford Wilson agreed.

The plantation houses were small, by the standards of the time, but they were immaculately clean. A small army of servants kept the entire area tidy, the pathways clear, the shrubbery cut, the grass clipped, and the fruit trees pruned. It was, Jose Palacios commented, the Garden of Eden. Even the weather was good. Sea breezes kept the place cool and evening showers dispelled the humidity. Their host, Don Joaquin de Mier, could not have been more gracious. He explained that he was, as they knew, a Spanish Royalist, but now that Colombia was firmly established as an independent country, he intended to stay and respect the laws and become a citizen, which was his right by birth. Surprisingly, he confessed openly to Bolivar's party that he had always greatly admired the Liberator even though he loathed his cause. "But now that I am a Colombian, just as you are, I can openly express my esteem. Only one such man as Don Simon Bolivar appears in a generation. He is exceptional. I am honored to have him as my guest."

"He might not be staying very long," said Fernando sadly.

Sitting beside the Liberator's bed, the French Doctor Reverend took Bolivar's pulse and listened to his heart. Jose Palacios stood nearby.

"You're doing fine. And this place is so clean. Do you enjoy the view, sir?" Doctor Reverend was exercising his best bedside manner with his most distinguished patient.

Bolivar didn't answer. Instead, he turned to Jose Palacios. "Let us go. Bring my luggage on board. They do not want us in this country. Let us go!"

Reverend looked at Palacios.

"He'll be all right in a moment, sir. He has these fits once in a while."

"He'll need a priest shortly." Reverend's remark was matter-of-fact. "Does he want one, do you know?"

"He's always lived by the rites and ceremonies of the Catholic Church. I know he'd want to die by them."

"I'll see that the Bishop of Santa Marta administers the last rites."

At that moment, Bolivar sat up in his bed and seemed to recognize Doctor Reverend for the first time. He smiled a greeting, then said, "Tell me, doctor, why did you come to America?"

Taken aback, Reverend answered, "Why, for the sake of liberty, sir."

"And have you found it here?"

"Certainly, Your Excellency."

"Ah," said Bolivar. "Then, you have been more fortunate than I."

The Liberator's friends sat with him. Laurencio Silva was holding a lit cigar, watching the smoke curl up to the ceiling. Bolivar coughed slightly. "Would you please move a little farther away, my friend." He motioned towards Silva.

Silva got up and pulled his chair several paces towards the door and then sat down again. He said, "The smell of tobacco never bothered you when Manuela was smoking it."

Bolivar's eyes lit up. "Ah, Manuela. Those were the days . . . yes, those were the days." And he smiled for the first time in a long while.

Clear-minded, Simon Bolivar looked around his room. He knew his end was approaching rapidly. He was unable to talk. His final pronouncements, made a few days before, kept running through his head.

Colombians, my last wishes are for the happiness of my country. I blush to admit that independence is the one good that we have achieved at the expense of everything else. Listen to my last words. At the moment when my life comes to an end, I implore and demand in the name of Colombia that you remain united. If my death can contribute anything toward the unification of the coun-

try I shall go to my grave in peace. I only wish I could have that consolation . . . Otherwise, I am the Liberator of a continent who, dying, regrets his achievement.

Slowly, he closed his eyes and a few minutes later, at one o'clock in the afternoon of December 17, 1830, Simon Bolivar drew his last labored breath.

In the middle of the crowd of mourners leaving the Cathedral of Santa Marta, Fernando Bolivar turned to General Silva. "Thank you for lending us your shirt. My uncle's clothes were in such bad shape, we really had nothing suitable to bury him in."

Silva, the tough old warrior, was fighting back tears. He still could not speak for a moment, but when he did, his words were clear. "Just think. The Liberator was buried in my shirt. I'm honored beyond any dreams of glory I could ever have conceived."

Silva was justified in feeling proud. Simon Bolivar was born rich. Perhaps the richest man in America. He led victorious armies and governed nations. He liberated an area five times larger than Europe, created half a dozen countries and became the hero of the western world. Honors and glory were heaped upon him, yet, with the exception of the expressions of appreciation from George Washington's family and from LaFayette, they meant nothing to him. Through it all, he remained truly noble. He wanted nothing for himself. On the contrary, during his life he gave so freely of all that he possessed that when he died his friends had to bury him in a borrowed shirt.

EPILOGUE

Had Simon Bolivar retired and departed for Europe after the victory of Ayacucho, nothing of consequence would have changed, except that the Great Bolivar would have emerged as a victorious general with his glory intact, his fame unsullied. Instead, he attempted to unite the countries he liberated and govern them democratically. In this, as we know, he failed, but at least he tried. It was this devotion to his cause and to his reputation at the expense of his personal happiness that best exemplifies the true essence of the man and helps explain his greatness.

For several years after his death, Bolivar's enemies continued to desecrate his memory, accusing him of all sorts of crimes against the common good. Then, twelve years after his death in exile, a transformation took place which can only be described as unique in world history. A convoy of ships filled with presidents of countries, foreign dignitaries, Princes of the Church and generals of armies transported his body in grandeur to Caracas, where he lay in state, honored and lauded extravagantly by some of the same men who had brought him down by their malevolent political hatred. To this day he lies venerated inside the Pantheon of Heroes, with an elite Honor Guard always present.

Cities, provinces and states are named Bolivar. An entire country, Bolivia, bears his name. His glory and prestige are unsurpassed. He has become more than a hero. To his countrymen, he is a demigod, a superman, the glorification of everything great in Hispanic America. Festivals and celebrations are held in his honor. He is extolled and glorified in speeches in senates and parliaments. His name is exalted in churches and schools.

Bolivar

There is not one person in all of South America who does not associate something great and overwhelming with Bolivar's name. Virtually every South American writer has written a lyrical work centering on Bolivar, the foremost theme of all Latin American literature. Books in their thousands are full of praise for his achievements. To the people of an entire continent, he is the greatest man who ever lived. And who is to say they are wrong?

POSTSCRIPTS

General Francisco de Paula Santander went into exile immediately after his conviction for complicity in the September 25th plot of 1828. He chose to go to Europe. From Hamburg, he wrote his version of the events that had taken place in Colombia. In 1831, the year after Bolivar's death, he was restored to all military honors and all rights of citizenship. As great an admirer of the United States as Bolivar, Santander returned from Europe via that country, where he visited New York and Philadelphia, Mount Vernon and West Point. He was in New York in 1832 when he received word that he had been elected provisional president of Colombia. He returned the middle of that year and, under a new constitution, was twice elected president. He married in 1836 and had three children. His presidency was notable for advances in public education and the signing of the final peace treaty with Spain. He died in 1840 of chronic hepatitis.

General Paez was elected president of Venezuela in 1831 and again in 1838. Illiterate as a young man, he learned to read and write when he was in his twenties. Later, he wrote extensively and ended up praising Simon Bolivar as belonging "among the band of modern men whose equals are to be found only when we reach back to the republican times of Greece and Rome . . ." He visited New York in 1850 and was received with full military honors. On a subsequent visit to New York in 1873, he died in that city and was given a stately military funeral by the troops from Governor's Island. There is a large oil portrait of General Paez in New York's City Hall.

* * *

General Daniel O'Leary was not with Bolivar when he died, but arrived in time to attend his funeral. He returned to Caracas in 1833. Afterwards, he served as an official in the Venezuelan Embassy in London. He was designated secretary of the diplomatic mission of Mariano Montilla to the courts of London, Paris and Madrid where they obtained recognition of Venezuela as an independent republic. He married Soledad Soublette, the sister of General Carlos Soublette, who later became president of Venezuela. In 1840, he again returned to Caracas and two years later participated in the ceremonies held when Bolivar's body was brought home. In 1844 he became British charge d'affaires and Consul General in Venezuela, where he wrote his famous memoirs, "Bolivar and the Wars of Independence." He was an admirer of Manuela and corresponded with her until his death. He wrote a volume about her and Bolivar, which was suppressed by the Venezuelan government, and was subsequently lost or destroyed. O'Leary died in 1854, and in 1875 his body was among the first to be placed in the Pantheon of Heroes in Caracas, where he rests beside his idol and hero, Simon Bolivar.

Simon Rodriguez served briefly as Director of Public Education in Bolivia. However, President Sucre had to bow to public will and dismiss Rodriguez, who shortly thereafter was forced to marry another young girl he had made pregnant. He went to a small town on the coast of Peru and ended his days in 1854, at the age of 85, teaching poor children without charge and making candles for a living. However, his epitaph could well read: "If it were not for Simon Rodriguez, the world would never have heard of Simon Bolivar".

After the Cordoba incident, Rupert Hand took retirement as a lieutenant colonel, with a lifetime pension merited by the almost mortal wounds he had received at Carabobo when he was a captain commanding the Third Company of the British Light Infantry Battalion. However, he returned to Venezuela in 1833, after Bolivar's death and the break up of Great Colombia, and became a teacher of English, the author of several texts, including "A Brief Analytical Explanation of the English Alphabet." He also became a member of the Municipal Council of Caracas, and in 1841 was the first professor of English at the University of Caracas. He died in that city in 1850.

General Rafael Urdaneta lived a respected and gentlemanly life until 1845, when he died of his chronic ailment of kidney stones while he was in Paris on a diplomatic mission for Venezuela. (He apologized to the archbishop who attended him for having the bad manners to die in his presence.) As the last president of Great Colombia, his place in history is secure. Had he lived, he would undoubtedly have been elected president of Venezuela. He had already declared his candidacy and was certain to have won the election.

Right after Bolivar died, Belford Wilson became desperately ill and was taken to Jamaica to recover. From there, he went back to England, but in 1832 he returned to South America as the British Consul in Lima, where he remained for several years. In 1842, he became charge d'affaires of Great Britain in Caracas, where he represented Her Britannic Majesty's Government with great distinction. In 1851, Colonel Wilson returned to London and was honored by Queen Victoria with a knighthood in The Most Honorable Order of the Bath. Sir Belford died peacefully in London in 1858.

General Jose Laurencio Silva, who was married to Simon Bolivar's niece, Felicia Bolivar y Tinoco, became commander in chief of the Venezuelan army in 1855 and remained in that position until he resigned in 1859. He then retired to private life and lived happily with his family in their home in Valencia and died in 1873 at the age of 82.

The heroic Colonel Juan Jose Rondon, who had won everlasting fame in countless battles, took part in repulsing the Spanish attempt to take Puerto Cabello in 1822. During the engagement, he received a slight wound in his heel from an enemy lance. Unfortunately, the wound became infected, and Rondon died of tetanus. Thus the dashing hero who turned the tide of battle at Vargas and fought so gallantly at Boyacá and Carabobo, died from an insignificant wound in an unimportant skirmish.

General Arthur Sandes married a girl from a prominent local family and settled down in Cuenca, Ecuador, where he was active in the Department of Public Instruction. (He was responsible for the founding of many schools throughout the province.) He died in 1832 at the age of 39.

Jose Palacios was liberally provided for in Simon Bolivar's will. He lived until 1842 and was able to savor the return of his hero to national glory. As if waiting for this event, he died in that same year.

The fiery Manuela Sáenz had been on the way to Santa Marta when the news of Bolivar's death reached her. It was said she fainted but upon recovering declared, "When he was alive, I loved Bolivar. Dead, I worship him!" Manuela was still young and beautiful; yet, as far as anyone knows, she remained faithful to Bolivar's memory and never took another lover. When the tough-minded Santander became president of Great Colombia, knowing her capacity for mischief as well as her implacable hatred of him, he exiled her. She ended up in the small port of Paita on Peru's pacific coast, where she made a living selling tobacco and translating documents at the American Consulate. Simon Rodriguez came to Paita especially to visit her, and, so great was her fame as Bolivar's consort, that the great Italian patriot, Garibaldi, also made the journey just to meet her and discuss his plans with her.

As Paita was the port of call for a large number of American whaling ships, her services as translator were in great demand. Once, after some of the crew of the whaler *Acushnet* mutinied, she had to take the testimony of a young, grey-eyed sailor. His easy banter appealed to her. The handsome young man and the still lovely matron became friends and mutual admirers. Before he left to sail across the Pacific, she called, "Wait. I have to write down your name, young man."

Over his shoulder he called back, "Melville, ma'am, Herman Melville."

One day in 1856, diphtheria came to Paita, brought in by a sick sailor. It carried away Manuela Sáenz. As was the practice, she was buried the day she died. When her old friend, General Antonio de la Guerra and his wife returned from her hasty funeral they found all her documents and letters from Bolivar, along with everything else she owned, burned to ashes as a precaution against the spread of the disease. Shifting through the debris, the old general picked up a piece of paper. It was charred around the edges, but the writing was clear and legible. He read it and, then, slowly lifted his eyes skyward, as his wife took the paper from his hand. She read, "Come. Come to me. Come now . . ."